Springer
Berlin
Heidelberg
New York
Hong Kong
London
Milan
Paris
Tokyo

Subhrajit Guhathakurta (Ed.)

Integrated Land Use and Environmental Models

A Survey of Current Applications and Research

With 96 Figures

Springer

Herberger Center for Design
Excellence
Arizona State University
Tempe Arizona

PROF. SUBHRAJIT GUHATHAKURTA
SCHOOL OF PLANNING AND LANDSCAPE
ARCHITECTURE
MAIL CODE 872005
ARIZONA STATE UNIVERSITY
TEMPE, AZ 85287-2005
USA
Subhro.guha@asu.edu

Published with the cooperation of:

Herberger Center for Design Excellence, College of Architecture and Environmental Design,
Arizona State University, Tempe, Arizona 85287-1905

Mary R. Kihl, Ph.D., AICP
Director/Associate Dean for Research

Julie A. Russ, Editor

ISBN 3-540-00576-5 Springer-Verlag Berlin Heidelberg New York

Cataloging-in-Publication Data applied for

A catalog record for this book is available from the Library of Congress.

Bibliographic information published by Die Deutsche Bibliothek
Die Deutsche Bibliothek lists this publication in the Deutsche Nationalbibliografie; detailed bibliographic data is available in the Internet at <http://dnb.ddb.de>.

Springer-Verlag Berlin Heidelberg New York
a member of BertelsmannSpringer Science+Business Media GmbH

http://www.springer.de

Printed in Germany

Camera ready by the authors
Cover design: E. Kirchner, Heidelberg
Cover image courtesy of Prof. Ramon J. Arrowsmith, Arizona State University
Printed on acid-free paper 30/2132/AO 5 4 3 2 1 0

Contents

Preface

This volume is the result of an invited symposium titled "Integrated Land-Use and Environmental Models: A Survey of Current Applications and Research" that was held in October 2000 at Arizona State University. The idea for the symposium arose from a belief held by many academics that we are at the watershed of a new generation of models that are more dynamic, more pragmatic, more interdisciplinary, and more amenable to collaborative decision making. Several academics and professionals engaged in urban research had long realized that domain-specific knowledge was inadequate for understanding and managing urban growth. While interdisciplinary approaches have become critical in most social research, one general area of knowledge that stands out as having the most wide-ranging impact on current urban modeling efforts is the field comprised of environmental sciences and ecology. The symposium offered a forum for academics and professionals engaged in urban and ecological modeling to exchange ideas and experiences, specifically in areas that overlapped urban and environmental issues. The contributions to this volume highlight the progress made in the various efforts to build integrated urban and environmental models. More importantly, each chapter shows how ideas have diffused across disciplinary boundaries to create better policy-relevant models. In addition, this book outlines some promising areas of research that could make important contributions to the field of urban and environmental modeling.

Integrated thinking about urban and environmental issues has been fundamental to the concept of sustainability. With the rapid growth of urban populations around the world, the demands on urban infrastructure are growing disproportionately faster than the rate at which such infrastructure is being expanded. The provision of urban infrastructure such as clean water, mass transportation, housing, open space, and waste removal is also a critical part of environmental management. Hence the environmental problems are compounded by both the increasing demands on natural resources by the growing urban populations and the inability to manage such population growth through the adequate provision of urban infrastructure. What is needed is a different approach to urban planning and environmental management that fundamentally changes the way cities have been designed and managed. Such an approach would also require innovative thinking about the connections between land, environment, human behavior, and human settlements. It would also necessitate new tools and better theories to understand the dynamics of urban growth and change. There is already a vast and growing literature on various aspects of urban-environment interactions. This book offers another dimension to this literature by compiling, in one volume, a set of theories and models that show how urban-environment interactions can be conceived, analyzed, and implemented in practice.

The thirteen chapters comprising this volume build on the general theme of changing perceptions about urban and environmental models by examining both the theoretical advances and current applications of such models. The chapters are not only products *for* the symposium but also products *of* the symposium given that most went through several revisions that reflect the influence of various conversations during and after the symposium. The papers have followed four interrelated themes that provide a natural organization of the sections in this volume. The first section is titled "Evolving Definitions—Changing Practices." In this section the fundamental shifts in urban modeling practices are examined in relation to the new theoretical and computational advances in the field. The second section, "Ecologic Processes and Their Land Use Implications," provides current examples of ecological models that influence land-use policy and planning. The next section on "Visualization, Representation, and Communication" deals exclusively with the science and art of geographic data generation and representation techniques. Finally, the section on "Socioeconomic Implications of Transportation and Land Use" examines the traditional domain of urban models from a sociological and environmental perspective and offers new insights on transportation planning.

No project of this scale can be undertaken without the active support and encouragement of a large number of individuals. There are many who have directly or indirectly helped in getting this project off the ground—more than I can do justice to in the space available in this book. The Herberger family provided the initial funding for the symposium as part of a gift to help initiate the Ph.D. program in Environmental Design and Planning at the College of Architecture and Environmental Design, Arizona State University. Besides this initial funding, I owe a debt to Professor David Pijawka, whose support and advise at the initial stages helped in expanding the scope and reach of this project. I am also grateful to Professor Frederick Steiner, the Director of the School of Planning and Landscape Architecture at Arizona State University at that time, who provided encouragement, advice, and momentum to the symposium through his active support and involvement. Additionally, the support provided by John Meunier, Dean of the College of Architecture and Environmental Design at Arizona State University at that time, and by Jonathan Fink, the Vice Provost for Research at Arizona State University has been crucial to the success of this endeavor. A significant partner in this project has been the Southwest Center for Environmental Research and Policy (SCERP). I gratefully acknowledge the support provided both in terms of funding and involvement by the SCERP management committee in general and Rick VanScoik in particular. I am also grateful to my colleagues at the School of Planning and Landscape Architecture at Arizona State University for their support, involvement, and insights.

The compilation of this book was a two-year project that involved a dedicated group of people, most importantly, the contributors to the volume. I thank each contributor for their continued engagement and effort in writing, revising and re-revising the manuscripts in light of conversations and editorial comments. I also thank each one of the editorial assistants for their excellent job in carefully editing and preparing the manuscripts for publications. They include, Jennifer Fraser, Audrey Morris, Vanessa Mallory, and particularly Julie Russ who finally managed to see this book to completion. Last, but not least, I wish to thank Mary Kihl, Director of the Herberger Center for Design Excellence, for her initiative in supporting the project in the face of many organizational and budgetary changes at the Center.

This book is dedicated to my students and to the students in the doctoral program in Environmental Design and Planning at Arizona State University.

Subhrajit Guhathakurta
School of Planning and Landscape Architecture
Arizona State University
February 2003

Section I: Evolving Definitions—Changing Practices

Advances in Urban and Environmental Modeling: Surveying the Terrain and Demarcating Frontiers

Subhrajit Guhathakurta

Introduction

While urban and environmental models have evolved in separate domains of knowledge, they have always shared a common goal, that is, to maintain and advance the relationship between Earth and its inhabitants. This goal is implicit in the specific issues that typically concern urban planners and environmental scientists. However, it has rarely been stated in such broad terms and has often fallen prey to competitive models of the economic marketplace. Planning, as seen from most perspectives, is a tool for more efficient exploitation of resources to meet the ever-expanding demand of human societies. This is achieved through building better organizational structures and accompanying institutions that are primarily geared to enable smooth operation of the market. In contrast, environmental scientists have repeatedly taken human societies to task for ignoring Earth's real capacity limits and for increasing the human footprint at the expense of other species. Until recently, urban models have concentrated on the consumption and use of land through better access and improved infrastructure. Environmental models, in contrast, have focused on sustainability of resource use and the maintenance of species' habitats. In essence, the objectives of the two modeling paradigms were at odds. The conflict is often characterized as the tussle between "brown" and "green" issues.

Over the past decade the forces bringing the two disciplines of environmental science and planning together have become stronger. There is now little doubt about the dialectical relationship between "brown" and "green" given that issues concerning one cannot be addressed without invoking the other. The environmental movement and the accompanying interest in sustainable development provided another major impetus for bridging the gap in knowledge domains. A number of international agreements such as Agenda 21, the Rio Declaration on Environment and Development, adopted in 1992, and the Earth Summit in 1995 pushed the cause of sustainable development across the world. Several nations followed with specific policy directives to implement some or most of the environmental agenda. For example, President Clinton created the President's Council on Sustainable Development and Canada developed its own Sustainable Development Agenda. Urban planners and environmental scientists have now begun to emphasize the importance of developing an integrated framework for modeling ecological and socioeconomic processes. This volume highlights some of the progress made in that effort and provides a roadmap for future research in integrated urban and environmental modeling.

Methodological Issues in Integrating Urban and Environmental Models

Although there is little argument against the rationale for integrating the two domains of knowledge, the methodological issues in the bridging process have posed some serious problems. Most urban models are still limited in their ability to address the environment. These models are rooted in economic theory and focus primarily on economic and spatial interaction among jobs, housing, and transportation. An economic framework has limitations in incorporating ecological dynamics since price signals play a marginal role in environmental processes. Similarly, until recently ecological modelers have concentrated on modeling species behavior in nonurban landscapes primarily by accounting for the flows of energy and matter through various natural systems. Also, urban and environmental modelers have distinctly different concepts of spatial and temporal processes. Spatial categories enter urban models at a high level of aggregation with limited interaction between them, hence are unsuitable for integration with ecological models at spatially detailed levels. In addition, the fascination with cross-sectional equilibrium in urban-economic models relegates temporal issues to the background. However, recent advances in the literature on agent-based models, system-dynamic processes, and complexity theory offer important insights about integrating social and ecological knowledge domains.

Agent-based processes examine the dynamic interaction between the choices made by various entities such as institutions, governments, businesses, and households. Ecosystem modelers have been using agent-based models to simulate population growth and decline as well as changes in environmental resource endowments. Typically, the components entered in an agent-based model interact with each other in the form of feedback processes. Such feedback processes can be negative or positive. Negative feedback from one component in the model leads to a response in other components that counteract the original change. Positive feedback does the opposite—it evokes a response from other components of the model that strengthens the original change. The interplay between the negative and positive feedback processes lead to the dynamic characteristics of the system being modeled.

The system-dynamic approach is ideally suited for modeling agent-based processes. It is also well adept at capturing emergent processes that exhibit complexity. Complexity is often manifested from simple rules applied to local phenomena such that aggregate patterns are clearly distinct from local behavior. Complexity studies now contend that detailed micro level studies and their dynamic properties are essential to understanding macro behavior. This is in contrast to the reductionist perspective that assumes that simpler, local level characteristics can be disaggregated from macro processes. The systems approach provides an elegant means of observing complexity that is emerging from simple rules of expected behavior.

Although system-dynamic approaches have been conceived as a "grand" approach that attempts to tie together multiple domains of knowledge, it cannot be expected to integrate different theoretical and epistemological domains that have framed disciplinary advances. System-dynamics is also limited in constructing theories since its principle purpose is to clarify, test, and unify a priori theoretical insights or "mental models." It is, therefore, a tool to refine and develop existing theories and extract insights about these theories as they play out in the real world. In addition, systems models lose their simplicity and elegance when spatial aspects of a system are included. The amount of computation increases exponentially with increasing resolution of spatial categories. However, spatially disaggregated dynamic-system models are being developed and tested. Some current and ongoing projects of such spatial models include the Spatial Modeling Environment being developed at the University of Maryland (Costanza et al. 1995; Voinov et al. 1999) and UrbanSim, a project housed in the University of Washington (Waddel 2002).

While there are several models available to address various aspects of the environment and economy, any one model would be too limited to capture this larger system. What is needed is a computing interface in which various models can be linked at spatial, temporal, and functional frames. The use of multiple models that interface with one another has several advantages. First, this approach would require fewer compromises and preserve, to a large extent, the integrity of the submodels. Second, it would allow greater examination of the model substructures and hence facilitate more vigorous discussions about the different epistemologies guiding model development. Third, this model would not be "owned" by any knowledge domain and would likely be truly interdisciplinary. Fourth, the use of multiple models would require a more conscious examination of the embedded narratives and allow the construction of a coherent plot. Thus the overarching narrative that weaves the model together serves as the glue for integrating different approaches.

Dilemma of Integration

The optimal extent of integration of multiple knowledge domains in a modeling framework has posed some theoretical and epistemological issues. The process of stitching together different theoretical frames inevitably entails some compromises and limitations. For example, levels of data accuracy required for some models may be unreasonable when applied to other models that work with heuristic approaches. Also the resolution of the data in space-time may be different for different aspects of the modeling exercise. In such cases, common frames need to be designed that can incorporate different approaches without compromising the integrity and elegance of individual model elements. More importantly, the ultimate goal of the modeling exercise has to be clearly defined, which can then guide the priorities structured within the model framework. Therefore, the model needs to reflect the goals as designed by the community of model users, rather than being just an academic showcase of modelers' individual or collective expertise.

Structuring the primary questions for the modeling exercise is a nontrivial activity. Especially in situations where "facts are uncertain, values in dispute, stakes high, and decisions urgent" (Funtowicz and Ravetz 1993: 744), conceptualizing the proper questions and casting the questions in the proper perspective can be extremely difficult. Even where there is agreement on broadly defined goals such as "economic justice" or "sustainability," operationalizing such goals within local realities is often unachievable due to ambiguity in interpreting these terms. Hence, a critical step in the modeling process requires the transformation of generic goals, usually encompassing an infinite information space, to a bounded information space serving specific objectives. This process sets into motion a series of uncertainties related to the selection of appropriate goals and assumptions that may not be well defined within the various disciplinary perspectives. This type of uncertainty is also known as the "problem structuring uncertainty" (Mayumi and Giampietro 2001).

A fundamental problem of urban and environmental models is their inability to conform to the strict rules of accuracy and testing that underlie physical models. According to Georgescu-Roegen (1971) in natural sciences "a model must be accurate in relation to the sharpest measuring instrument [and] there is an objective sense" in comparing the results of various formal systems of a physical model. Social scientists can rarely aspire to the level of formal testing that a natural scientist is familiar with. Hence, socioeconomic models are only "similes" that offer insights and guidance for decision making. Also, representations of shared perceptions require a level of sensitivity and transparency that elevate the modeling exercise to a skill that is acquired less formally. With the emergent awareness of the politics and limitations of expertise, particularly with the inability of experts to offer certainty or control, the credibility of models and modeling exercises have become eroded.

Therefore, at the root of the issue of model integration is not the methodological problem of integrating spatial and temporal scales, or even the formulation of a bounded information space within which to construct an integrated model or interacting models. The foundational issue is the issue of trust. The issue of trust emerges at various levels, from the disciplinary suspicions of the motivations and hidden agendas of other disciplines represented in the team, to the stakeholders' skepticism of technical experts who often disregard critical local knowledge in favor of formal knowledge. In addition, it is not unknown for experts to be skeptical of tacit or local knowledge and, in the process, assume a sense of arrogance about their own expertise. Often ensuring that experts are sensitive and responsive to lay people is more problematic than developing a sense of confidence among the stakeholders associated with the modeling process.

In the context of planning, the distribution and management of risk and reduction of uncertainty have always been an important objective of planners. Given that most situations in the social sphere have attributes of reflexivity and adaptability, they are difficult to predict with a high degree of certainty. However in the real world, decisions are routinely made under circumstances where knowledge and information are sparse. Models help in limiting the risks of decision making under less-than-ideal conditions through intelligent speculative inquiry. Such models are only "similes" that focus attention on some attributes of our social and ecological environment and sharpen our understanding of the social order. Therefore they should only be treated as learning tools, and not predictive tools. In so far as an integrated model allows us to explore the various interactions between the parts modeled and allows debate/discussion to influence it, this model would serve a useful purpose. However, if the integrated model only serves to generate results within an opaque, "black-box" environment, it would fail to deliver any benefits from the modeling exercise.

Making Models Reflexive and Pedagogic

Reflexivity means "the application of a theory's assumptions to the theory itself, or, more broadly, the self-monitoring of an expert system, in which the latter questions itself according to its own assumptions" (Lash and Urry 1994, 5). Reflexive models are models that go beyond the accepted theory and formal methodologies of traditional research to enable the evaluation of a multiplicity of perspectives, which are derived from both formal and informal sources. In so doing, the models are themselves interactively transformed to accommodate each new information or perspective. Such models are also pedagogic in the sense that the objective of modeling is not to predict and control, but to learn about actions, reactions, causes, and consequences of social and ecological processes. Reflexive and pedagogic models have a dual nature: 1) they are intrinsically dialogic given that the process moves through debate and dialog and engenders further discussion beyond the model boundaries; and 2) they provide a means of dealing with, as well as adapting to, complexity.

Reflexivity ensures that knowledge is informed by praxis and vice versa. Increasingly the relevance of science is becoming ever more crucial, but at the same time less sufficient, in uncovering social truths. Hence as long as discourse about science and scientific modeling remains limited to experts, the process will simply reinforce the underlying biases and continue to shut out the large community of peer groups and stakeholders. There is a need for a new organizing principle for integrated modeling that is dynamic, systemic, and pragmatic. While there are no standard methods for building reflexive and pedagogic models, some overarching guiding principles can be distilled from critical inquiry and earlier experience.

Undoubtedly these guiding principles do not provide easy solutions and can be hairy to implement, especially when team dynamics become problematic. Regardless, keeping the guiding principles in focus opens up several pathways to building useful models of social behavior, its causes, and implications.

First, building reflexive models requires a commitment from all participants to focus on an iterative open process rather than an activity happening at a limited point in space and time. This becomes a major sticking point especially since most modeling efforts have limited resources and strict deliverables at various stages. Given the shift in focus in this approach from the technical aspects to the social aspects of modeling, such onerous expectations can be stifling and counterproductive. However, some benchmarks are needed to measure progress of the modeling process. The outcome measures that work best are those that relate to decisions made along the way rather than to the number and specifics of the tasks accomplished.

Second, among the decisions made early in the modeling project, perhaps the most critical would involve the definitions and formal identities of relevant ecological subsystems that would form the core of the model. Such definitions would rely on perspectives from various stakeholders and participants. Also, the institutional and political settings, technological possibilities and limitations, as well as cultural traditions would come into play in solidifying the subsystem definitions. This process would entail a movement from generic goals that are easy to articulate to more specific, often contentious, objectives. While the formulation of clear definitions and decisions on a specific system boundary are necessary steps, a reflexive process must also be amenable to change if at some point the definitions are reshaped requiring a restructuring of relevant subsystems. Hence the initial agreement on an iterative, open process is important.

Third, it is important to pay some attention to selecting appropriate criteria for judging the performance of the model. These criteria should relate to the objectives defined in the second step. However, given the multidimensionality of social and ecological processes, any selected range of indicators based on specific objectives can provide a complex and conflicting picture. Hence, it may be difficult to obtain an unambiguous picture without debating the relative weights on various objectives defined within the modeling scope. The model itself needs to be transparent in the way it provides information about different aspects being modeled. An interactive model in which the relative weights on indicators can be changed to show different possibilities is ideal in this context. Such a model would facilitate discussion of priorities and focus attention on those objectives that are critical and others that may need attention at a later date.

Fourth, nitty-gritty technical issues need to be resolved before the model is constructed. These issues include the choice of theoretical models, measurement schemes and analysis scales, data collection methods, desired accuracy of information, and the space-time horizon to be modeled. Often such technical issues limit the boundary conditions of the model, which in turn leads to a model that does not address all defined objectives. Although efficacy dictates the use of familiar, albeit limited, models, this may defeat the purpose of developing the model itself. Hence concerted effort is needed to address the technical issues by employing knowledge base developed in multiple disciplines. A multidisciplinary approach provides a rich base of theory and methods that apply across various contexts with minimal modifications.

Finally, there is a need to maintain a level of humility among the participants of the modeling exercise since social and ecological models are inherently gross simplifications of reality. Most complex systems are also self-modifying systems that cannot be modeled with a high level of certainty over long periods. As Mandlebrot (1982) has shown, such systems also exhibit nested hierarchies, making them indeterminate across scales. Therefore, a certain degree of uncertainty is unavoidable. It is important to keep in mind the pedagogic nature of the exercise and focus on the learning process, which may provide clues about dealing with complex issues on the ground. To expect unambiguous answers about specific tasks from the model is unreasonable. Even if the modeling exercise raises interesting questions instead of providing (often incorrect) answers, it would have made an important contribution to our knowledge.

Constructing Narratives from Simulation Models

Simulation models have several useful properties that allow effective communication, discussion, and learning. Many simulations involve physical models that are scaled down replicas of the original processes. Examples of such physical models include wind tunnel experiments in aviation, automobile crash tests using dummies, and other complex natural processes like the physical model of the Mississippi River that the Army Corp of Engineers uses to study the impacts of flooding. Over the last three decades computer simulations have slowly replaced many physical models and have allowed several other forms of simulations to be constructed relatively cheaply. Computer simulations have revolutionized meteorological studies, aviation and missile technology, design and development of nuclear reactors, and the study of environmental change.

When testing simulation models, emergent properties can be observed that have not yet been analytically described. In fact, many emergent properties are difficult to describe analytically with a high degree of precision. However, by repeated observations under test conditions, an appreciation of the resultant effects can be gained that may either provide valuable information for decision making or illuminate certain properties requiring further scrutiny. In such cases the computer simulation bridges the gap between "speculative inquiry," a domain of philosophy, and the empirical techniques that have dominated scientific research over the past two centuries. The computer program serves as an analogy of an explanatory theory, which is tested within the controlled "virtual" environment and modified if the results do not conform to observed facts. The testing of simulation models may continue through a structured analytical process or through an unstructured iterative (sometimes numerical) process or both. The objective is to express, refine, and test the underlying explanatory theory under specific contexts, which may lead to the application of the simulation model within this context for decision making.

Another important feature of simulation models is that they explicitly include the element of time. Although real world processes evolve dynamically over time, most urban models have failed to incorporate this important element other than as an explanatory variable among many others. This is especially true of urban models based on economic theory. In contrast, a simulation model unfolds over time, hence capturing, in a compressed form, the passage of time during the evolution of the process being modeled. By structuring a process as a sequence of events, decisions, and circumstances, a simulation model offers the possibility of describing the model in a narrative form. If, as suggested by Aristotle, a narrative is a representation of events, circumstances, and processes presented by a narrator, then simply running a simulation model does not constitute a narrative. However, an observed simulation result that is described logically by a narrator to construct a meaningful story would indeed constitute a narrative. The progression of a narrative is selective because the events are chosen and structured by individuals specifically to suggest a coherent plot. A narrative based on a simulation is therefore intersubjective as well as communicative since the plot renders meaning to specific experiences or logical deductions. The simulation narrative is also fundamentally different from a novel or a drama. In the words of Janet Murray, "whereas novels allow us to explore character and drama allows us to explore action, simulation narrative can allow us to explore process. Because the computer is a procedural medium, it does not describe or observe behavioral patterns, the way printed text or moving photography does; it embodies and executes them" (1993, 181).

Acknowledging the narrative aspects of simulation models allows for a significant switch in our cognitive perception from the "paradigmatic" to the "narrative." According to Bruner (1986), the "paradigmatic" realm is the world of abstract and general theories that are empirically verified in the objective world. In contrast, the "narrative" mode of thought focuses on particular events and experiences over time that gain credence through their lifelikeness. It is the quality of meaningfulness rather than factual accuracy that renders a narrative credible. Rendering meaning to a simulation model is as much related to an act of interpretation as is communicating a story because meaning does not preceed the interpretation of experience. Concepts such as "explanation," "validity," and "verification" are redefined in the narrative forms of inquiry. The search is not for mathematical certainty but for results that are believable, meaningful, and verisimilar.

Projecting the Trends in Urban and Environmental Modeling

The traditions of urban and environmental modeling have now begun to move in a common direction. The growing interaction between these two academic domains has indeed improved our understanding of social and ecological relationships. More importantly, it has helped in developing an appreciation among the modeling community of the inherent uncertainties of complex adaptive systems and the inadequacy of most existing tools in uncovering the intricate dynamics of social and ecological processes. The current approach acknowledges that the multidimensional character of social and ecological systems necessitates a multidisciplinary approach to modeling. Most current modeling projects in the United States and abroad relies on multidisciplinary teams. This has allowed the possibility of particularistic knowledge domains to address similar problems outside of that domain in a different context. The social sciences are replete with examples of theories that have filtered down from other disciplines. Some well known examples include the "gravity model" used mostly in transportation forecasting and based on Newtonian Physics and "social ecology" models based on Darwinism. Currently, optimization of network signal flows, a standard and well-known process in chip design, has been shown to have significant application in transportation planning (Tayal 2001). Multidisciplinary approaches have improved the pace and efficiency of knowledge diffusion across knowledge domains. Current trends indicate that the use of multidisciplinary teams in building urban and environmental models will be the norm rather than the exception.

The complex-adaptive character of social and ecological problems will require a reevaluation of model purpose and function. Models will tend to be primarily pedagogic tools for learning, communication, and decision making. This pedagogic aspect of models is facilitated by a modular approach that uses a common framework to selectively and interactively bring together a set of sub models. Therefore, a modeling environment, within which models are constructed from a modular toolkit and other helper applications, would replace the large, integrated model. This modeling environment would provide a basis for critical inquiry that informs and is informed by the modeling exercise. Given that dynamic simulations and visualization offer several advantages—such as (1) the ability to address unstructured problems, (2) the possibility of visualizing emergent properties that are often unexpected, and (3) the ability to communicate through narratives—it is likely that most urban and environmental models would include some aspect of simulation. Shifting the emphasis from empirical/deterministic models to simulation leads to a new form of expression that may offer a different understanding of the social and ecological evolution.

References

Bruner, J. 1986. *Actual minds, possible worlds*. Cambridge, Mass.: Harvard University Press.

Costanza, R., L. Wainger, and N. Bockstael. 1995. Integrated ecological economic systems modeling: Theoretical issues and practical applications. In *Integrating economic and ecological indicators*, edited by J. W. Milon and J. F. Shogren. Westport, Conn.: Praeger.

Funtowicz, S. O. and J. R. Ravetz. 1993. Science for the post-normal age. *Futures* 25:735–755.

Georgescu-Roegen, N. 1971. *The entropy law and the economic process*. Cambridge, Mass.: Harvard University Press.

Lash, S. and J. Urry. 1994. *Economies of signs and space*. Thousand Oaks, Calif.: Sage Publications.

Mandelbrot, B. B. 1982. *The fractal geometry of nature*. San Francisco: W. H. Freeman.

Mayumi, K. and M. Giampietro. 2001. The epistemological challenge of modeling sustainability: Risk, uncertainty and ignorance. Paper prepared for *Frontiers 1*. Cambridge, UK, July 4–7.

Murray, J. H. 1993. *Hamlet on the holodeck: The future of narrative in cyberspace*. New York: Free Press.

Tayal, T. 2001. *Optimization of network alignment for light rail transit: Phoenix*. Master of Environmental Planning Thesis, Arizona State University.

Voinov A, R. Costanza, L. Wainger, R. Boumans, F. Villa, T. Maxwell, and H. Voinov. 1999. Patuxent landscape model: Integrated ecological economic modeling of a watershed. *Environmental Modelling and Software* 14 (5):473–491.

Waddel, P. A. 2002. UrbanSim: Modeling urban development for land use, transportation and environmental planning. *Journal of the American Planning Association* 68 (3):297–314.

New Developments in Urban Modeling: Simulation, Representation, and Visualization

Michael Batty

What Are Models? What Is Modeling?

Ludwig Wittgenstein's early definition "a model is a picture of reality" (1921, 8) suggests that the basic idea of a model is rooted in the philosophy of science and in scientific method. Yet it was not until the 1950s that the idea of a model began to be used widely in science, and then the classic definition of model was as "a simplification" rather than "a picture of reality" (Lowry 1965). This was the meaning ascribed to its use in the 1960s as the idea of a model gathered pace as a vehicle on which to develop good theory and applications in countless areas of the physical and social sciences. The term came into fashion first in North America where faith in science was officially translated into various large projects in defense, space exploration, and business. Advances in computing also accelerated the notion that models could be actually built and operated to make better predictions and even better designs for a variety of complex systems.

Before mid-century, the term model was used for its traditional purpose, which reflected scaled-down versions of the real thing which were, of course, simplifications but in a literal, physical sense. The term did, however, begin to enter the academic lexicon as scientists and social scientists began to think more formally about complex systems. In the post-war years, the word seemed to conjure up the power of science and technology in providing tools for understanding less glamorous but equally complex domains, especially in the human and policy sciences. The idea that complex systems might be modeled, hence controlled and thence designed or restructured came to symbolize the cutting edge of quantitative social science and complex systems theory in business, defense, and government. Indeed, as good a symbol of the imagery of these times as any is contained in the titles of the various works of the eminent economist-psychologist Herbert Simon whose books were successively called: *Models of Man* (1957), *Models of Discovery* (1977), *Models of Thought* (1979 and 1989), *Models of Bounded Rationality* (1982), and finally his autobiography published in 1991 titled *Models of My Life*.

Since the 1970s, the term has become widespread, being used to describe many different types of human process or operation, from the most abstract to the most routine. Indeed as its use has broadened, its power to hold all before it has lessened while the term itself has come to describe the widest possible range of simplifications or pictures of reality, as well as prototypical applications and inventions. This trend can be seen quite clearly in urban and regional planning. For example, fewer and fewer books are being written using the title Model, while the term itself is being used increasingly in everyday professional language. In fact, even when it first came to be used generally in a technical sense within planning, its usage was broadly based. In an article characteristic of our concern for definition some 30 years ago, Echenique says, "A model is a representation of a reality, in which the representation is made by the expression of certain relevant characteristics of the observed reality and where the reality consists of objects or systems that exist, have existed, or may exist"(1972, 164). His focus on what constitutes the reality as well as the model broadens this to include past as well as future systems, and as the future is unknowable, this use of models extends from the factual to the fictional, from science to design.

In this paper, our concern is not merely to recount the history of urban modeling for this has been done many times before (Harris 1968; Batty 1979; Batty 1994; Wegener 1994; Wilson 1998), but to review this history in terms of the changing perceptions of what models in planning are for, what we might expect of them, and how attitudes and practices continue to change with respect to their use. In this sense, we will exploit four themes that characterize the field. One of these—the changing significance of the term model—we have already noted, but a second theme involves the way those using models in planning have become more comfortable with abstractions, perhaps even theory, in thinking about the planning task. A third theme involves the changing role of data which, 50 years ago, was rarely thought about as being any form of abstraction, while a fourth theme involves the extent to which not only planners but their publics and clients might be involved in using abstract tools to inform their concerns and interests. These four themes—model use, the role of abstract thinking, data as models, and participation using models—involve ideas that we will weave into our historical analysis so that we might provide a balanced perspective on the wider role of models in planning.

This paper is organized in a straightforward way, beginning with a brief summary of the way models were first introduced into planning in the post-war years. The high point was reached in the 1960s and we will argue that, by then, most of the key ideas that have dominated their practical application had been introduced. We will then recount the way the field withdrew into itself, picking up on the way the computer revolution spurred new developments in representation (but rarely in analysis and simulation), culminating in a concern for manipulating data for more pragmatic and less ambitious ends than had been assumed a generation or more before. As the computer revolution has continued, the quest for better and better digital representation has grown rapidly and currently is moving fast toward representing systems more realistically, thus forcing the field once again to reconsider its roots in the physical representation of cities. New kinds of abstraction have merged which now link the digital to the real world through various interfaces which are not only opening up these kinds of representation to professionals but also to a wider public through networked communications. At the same time, new kinds of mathematical models of cities are being developed, very different from those that marked the field in the 1960s for the emphasis now is more on pedagogic use. A new pluralism dominates the field. The planner's tool box is much expanded as models emphasizing representation as well as process and design characterize the field. In conclusion, we will briefly draw all these themes together, attempting a synthesis and some speculation.

In the Beginning: Land-use and Transportation Modeling to the 1960s

As we have implied already, the term model within planning conjures up the popular image of a scale model of the urban environment, extensively used in architecture and urban design, and traditionally produced from tangible materials where the emphasis is on visual appearance. In a way, this usage still remains at the heart of the mainstream, at least in the public's imagination. Indeed, one of our themes is that we have now come full circle with the same features of the city being represented no longer with tangible materials but with ethereal ones. Our icons are no longer scaled-down versions of what we see manufactured from traditional materials but digital versions of the same—digital toys—the toys of tomorrow as they have been called by researchers at MIT's Media Lab. Traditional iconic models go back to prehistory but the change in usage in urban planning can be traced to the immediate post-war years. As noted, the idea of abstract mathematical models of cities and their functioning was rooted to developments in systems theory, mathematics within economics, social physics, and much else in the interwar years, but the shift in thinking about cities which presages such applications did not begin until the 1950s.

The first mathematical models to emerge in North America in these years were associated with the beginnings of transportation studies, and these were quickly extended to land use and its prediction. In the 1950s, one of the watchwords of the urban planner was that "traffic is a function of land use," and this message was to dominate the wave of land use–transport studies that began in the wake of the massive federal highway building program that began during those years. Two completely separate but nevertheless essential elements made such modeling possible. First, for more than 50 years, a succession of researchers had fashioned a series of rudimentary theories about how urban activities located in cities. These ideas emerged from economics as location theory but were complemented by the application of classical physics to geographical problems—social physics—all of which came to be tied up in the burgeoning field of regional science, which provided the early intellectual foundations. The second element related to computation. Almost as soon as digital computers were invented, their prime focus was to provide vehicles for intensive computation in science and then commerce. Cities and their transport systems provided prime candidates for these new technologies.

The final ingredient that set the world of planning humming in the 1960s was the policy context. Cities were growing and restructuring as populations became richer and more mobile, while problems of deprivation and renewal took on a new urgency. Models looked like part of the answer to get a handle on such complexity. The story of these years is well known. A flurry of different modeling styles emerged in the late 1950s and several applications were made in the subsequent decade. The effort was overly ambitious in many ways. Data constituted a problem and many efforts became morassed in data collection. Several models remained incomplete because computation was expensive and problematic. All were adversely affected by budget constraints, and the organizational management of such efforts was poorly conceived, but the singly biggest problems were that the models were not attuned to what policy makers wanted (Brewer 1973). In parallel, the theory on which such models were built was inevitably crude, mirroring our poor understanding of how cities worked. Many models were unintelligible to anyone but their developers and produced outputs that were often fanciful in their implications. Lee (1973) summed it all up in his paper "Requiem for Large-Scale Models."

Problems of data remain today, although they are changing while the computation issue has all but been solved. However, the real issue related then as now to what was being modeled. In essence, cities were conceived of as being in equilibrium and thus these models attempted to simulate how activities—land uses—located with respect to one another at a snapshot in time. Distance or its generalization as accessibility held the key to such spatial interdependence. If one could develop formal relationships between activities in space based on distance—the key organizing concept in social physics and transportation modeling—then models could be calibrated to reproduce the existing situation, and thence used in comparative static fashion, to make one-shot predictions of what the equilibrium of a future state of the city might look like. All the models that were produced during that time made this assumption. Few if any attempted to model the dynamics of urban change for it was assumed that cities were most often in or at least near equilibrium and thus even if their dynamics were modeled, these would simply return the city to a static spatial equilibrium. After all, cities had looked pretty much the same for 100 years or more, they just seemed to get bigger with no real differences in kind.

Linear econometric models such as the EMPIRIC formed a class of models in which spatial interaction was implicit (Hill 1965), in contrast to those non-linear structures that attempted to model spatial interaction explicitly, such as Lowry's (1964) Pittsburgh model. Models based not on simulating the existing city system but on optimizing it according to various predetermined planning goals were also attempted as were more behaviorally based models, which linked economic optimizing to location decisions (Batty 1972). All of these models operated at a fairly aggregate level in that census tracts and traffic zones formed the level at which cities were represented. This immediately introduced a level of abstraction into the process that took these models away from physical design. In short, these models were aggregative, static, physical, and spatial in focus, simplistic in the way they treated urban behavior, but rather abstract. Thus they were hard to relate to structures on the ground, which made them remote from those interested and entrusted in making decisions about the future form of cities.

From our current vantage point of the millennium, this early effort seemed doomed given the conditions under which such models were built. Their failure to address policy issues directly was their downfall, but the real problem resided in our ignorance of how cities actually worked. We will return to this later, but for the moment it is important to note how this early experience conditioned later developments. Three distinct approaches emerged from these models, each of which complements and has influenced the others to different degrees. The most obvious was involved in the extension of these models to embrace other sectors. These models came to be nested in wider spatial and aspatial economic and demographic structures, which enabled populations and employments to be handled in a more integrated way. As part of this, some of these models were generalized to operating over discrete time intervals but their structure is little different from those of the 1960s. They are still static and aggregative in form, notwithstanding the fact that they can now deal with many sectors and many other kinds of spatial flow (Wilson 1974; Batty 1976).

The second type of model involved true disaggregation of the spatial units defining cities into individuals. Developments in transport modeling around the idea of discrete choice, which is strongly linked to economic preference theory, are central to this. Such models, although still largely static, are able to treat space and time as attributes of the choice process, and thus have considerable potential for supporting contemporary views of the city as a system in perpetual disequilibrium. Such models allow much finer tuning of spatial behavior and have been embedded within wider model structures based on microsimulation, but their performance in mirroring real behavior has been worse than their aggregate cousins (Ben Akiva and Lerman 1985). The third type of model represents a synthesis between both the aggregative and disaggregative through the notion of economic optimization. Optimization had always been a theme in urban modeling, given the policy context of urban planning and the idea of providing some unification to the field by linking the way activities actually located to how they might best locate was an important quest. In many senses, this unification has actually been accomplished in particular through the linking of spatial interaction modeling with discrete choice theory in terms of entropy-utility maximizing, although few operational models have emerged from these insights (Wilson et al. 1981).

What is particularly important is that many of these approaches first developed in 1950s and 1960s are still highly significant 40 years on, and are a part of a new impetus in land use–transport modeling, at least within the United States. In fact, worldwide there are really only two serious practical efforts continuing the work from these early years, notwithstanding the many individuals who have continued to work in the field. These are the DRAM/EMPAL efforts of Putman at the University of Pennsylvania, which can be traced to a modification of Lowry's early Pittsburgh model and which has now been quite widely applied over the last 30 years to many cities in North America (Putman 1983, 1991). Then there is the Lowry-based MEPLAN model developed by Echenique at the University of

Cambridge, which has been widely applied in Europe and South America (Echenique 1994). Both these efforts have continued through spin-off consulting companies rather than in the university environments that nurtured them. Although both have embraced disaggregation and some individual behavioral simulation, they are still aggregate and static and suffer from these shortcomings, which is not to say that they are not useful for particular problem contexts. There are various efforts, such as the TRANUS model, which have spun off from the Echenique work (De La Barra 1989). The model systems developed by Wilson at Leeds, Kain at Harvard, and Brotchie at CSIRO, among others, though used for 20 years or more, had been abandoned by the early 1990s. Individual efforts such as Wegener's model at Dortmund and Anas's at Buffalo have continued, but these are one-off efforts without widespread application. New efforts such as Waddell's (2000) UrbanSim model are significant and we will return to these later. Wegener (1994) provides a very good summary of applications worldwide which, though nearly 10 years old, still provides a useful measure of where the action is.

Divergent Directions but a Reawakening of Interest

There are many commentaries on the first and subsequent waves of urban modeling up until the 1990s (see for example, the papers by Batty, Harris, Wegener, Klosterman, and Lee in the Journal of the American Planning Association 1994). All of these argue, to a greater or lesser extent, that although logistical issues were largely responsible for the failure of these urban modeling efforts to become institutionalized within urban policy making, the real difficulties lay in deeper concerns, in questions of theory and of policy relevance. The way cities were conceptualized as being in equilibrium; being focused around simple processes of movement, interaction, and location; and as being organized into homogeneous areas; have been thrown into grave doubt in the last 30 years as we have moved from industrial to post-industrial society and as the economy has become global. The way policies could be tested and explored in these models was also in doubt. Most of the models that became operational focused on a narrow range of issues from a policy standpoint—urban growth, transport infrastructure—but not on questions of spatial equity, redevelopment and renaissance, ghettoization, housing market analysis, and the like, which are intrinsically more difficult to associate with unambiguous spatial effects. In short, if anything, the first and later generations of operational models were too spatial, ignoring effects on other sectors and groups as well as being unable to deal with competition in time.

The disillusion with aggregate static models based on land use and transport was evident by the time Lee (1973) presented his critique. In particular, the inability to provide sectoral detail and to orient such models with explicit economic behavior was being challenged by the development of discrete choice theory. The failure of these models to deal with time and process, however, required a somewhat more fundamental shift. Apart from Jay Forrester's (1969) somewhat iconoclastic attempt to change the ground rules of urban modeling from space to time through his systems dynamics ideas (which considerably energized the field), ideas about how to treat time in ways other than simple linear progressions of cause and effect were being radically reviewed in science and in mathematics itself. The idea of discontinuities in smooth change through notions of catastrophe and chaos came onto the agenda during those years and various groups began to speculate as to how such ideas might account for spontaneous and radical change in cities, such as the growth of edge cities. Models based on structures that allowed bifurcations in growth paths, enabling surprising or novel change to take place, were developed at different scales by Wilson (1981) in the spatial interaction paradigm, Allen (1997) through ideas from non-equilibrium physics, and Dendrinos (1992) through ecological models and chaos theory. But in all these cases, the models were not conceived to be operational in the way of their predecessor models of the 1960s.

This was as much because noise and variation constitutes a major element within the processes which such models simulate, and hence deterministic outcomes are not the main focus. In the last decade, a further twist has occurred as the focus has moved back to ideas in social physics, in scaling, and in the far-from-equilibrium processes that such signatures imply. This is bound up with yet further disaggregation of urban processes and states to the point where individual agents and actors are being simulated. Such models focus on the way local actions generate global outcomes, on the way system properties emerge from the bottom up, and on the way systems maintain critical levels. Such is the essence of complexity theory and although many different kinds of urban model are currently being explored, few if any of these have operational content or focus as a brief review of the literature associated with these kinds of efforts reveals (see, for example, Batty and Longley 1994; Portugali 2000).

In the 1970s the demand for operational urban models did decline, though by the late 1980s the need for systematic frameworks for dealing with land use–transport interactions was evident once again. Federal mandates in the United States required municipalities to assess the impact of pollution generated by new road building on local communities, while the problem of sprawl set against new ecological concerns raised the need for some large-scale systematic assessment and predictions of the impacts of urban growth. Traditional urban models are once again being seriously considered for application with the DRAM/EMPAL and MEPLAN frameworks still at the core of such applications. In Europe, the MEPLAN, TRANUS, and Dortmund models are being made consistent with one another through various Europe-wide projects in terms of data and output media and their link to evaluation processes. In North America, other modeling efforts such as Landis's (1994) California Urban Futures Model (CUF), Waddell's (2000) UrbanSim, and a series of variants that build on these are being applied (Schock 2000). Most of these types of model fall within the traditions established through regional science and transportation modeling back in the 1950s and 1960s but, interestingly, there are other traditions emerging that seek to provide a rather different focus on urban growth, geared more to representation than to urban process, to physical growth rather than economic structure. To provide some completeness to our discussion, we need to identify these.

In the 1960s, the ideas that land use generated traffic and that any model must track such interactions were central to most of the models then built. But there were significant exceptions, in particular Chapin's effort in North Carolina, in which the emphasis was on the land development process as a driver of urban growth. The models that Chapin and his colleagues proposed (Chapin and Weiss 1968) focused upon the way growth took place through time, growth being a function of the physical infrastructure and accessibility within different parts of the city. This concern for modeling growth through time resonates strongly with current concerns involving urban sprawl. In some senses, Landis's (1994) CUF model is close in spirit to Chapin's original efforts, although a much closer parallel is in the spate of urban growth models based on cellular automata (CA) ideas which have suddenly mushroomed in North America through the efforts of the U.S. Geological Survey to come to grips with urban dynamics and sprawl (Schock 2000). These models are also much closer to ideas that are being developed for new ways of treating dynamics in cities and to notions about cities in terms of their physical morphology. They generally do not contain any processes that incorporate spatial interaction, in that urban growth is simulated as a process of diffusion with various types of infrastructure both driving and being driven by the growth that takes place. Insofar as there are explicitly dynamical processes at work in such models, these are ones of local diffusion in the presence of various constraints and noise.

In many ways, these urban growth models are much more pragmatically structured than their traditional counterparts based on spatial interaction. In fact, their appeal to CA ideas is largely incidental. They rarely invoke the strict limits of CA, and hence they are better termed cell-space or grid-space models. One of their main advantages is that they are closely consistent with the functionality and data associated with contemporary GIS, particularly raster-based data such as satellite imagery, which provides a superb backdrop for a study of the dynamics of urban sprawl. But such models have not been developed in the depth associated with more traditional models, and there are clear dangers in developing such untested and untried models for operational policy purposes. Nevertheless, it is worth illustrating the state of the art at this time with one such model in this newer tradition. Batty, Xie, and Sun (1999) are developing such a model for the town of Ann Arbor, Michigan, where there has been substantial urban sprawl over the last 20 years. One of the features of this model is that it makes use of various mechanisms involving diffusion and capacity constraints as well as differential interaction fields around cells of development, although spatial interactions are not computed explicitly. It takes data from desktop GIS packages such as ESRI's ArcView and simulates growth (and decline) starting from a set of seed sites consistent with urban development between two points in time. Figure 1 shows how the model is able to generate growth consistent with capacity constraints, thus illustrating how the model's dynamics work. Figure 2 show various scenarios for growth between 1990 and 1995 from the set of seed sites which form the development from 1985 to 1990.

In a paper such as this, it is not possible to present how any of the models actually work, though Figures 1 and 2 illustrate how far we have come in the last 40 years in developing software and data that can be used for widespread exploration of new model forms. Moreover, the software that we have developed is characteristic of the shift from simulation to representation that has occurred in this field in the last 20 years, as there has been a sea change towards digital representation through graphics and GIS. In the next section, we will examine this revolution in spatial representation but suffice it to say that in the millennium, the mathematical models of cities that we currently have at our disposal are not good, they are inconsistent with respect to what we know about cities and what we require of urban policy while those that have been explored most represent a way of thinking which is no longer popular. A particularly timely review of the field with respect to operational modeling has recently been provided by Schock (2000) for the Environmental Protection Agency, but it is hard not to draw the conclusion that more modest, data-driven approaches based on GIS might in many ways be preferable for urban policy analysis.

Figure 1. Typical urban growth model based local diffusion using CA.

1.1 A hypothetical city is planted in the middle of the space.

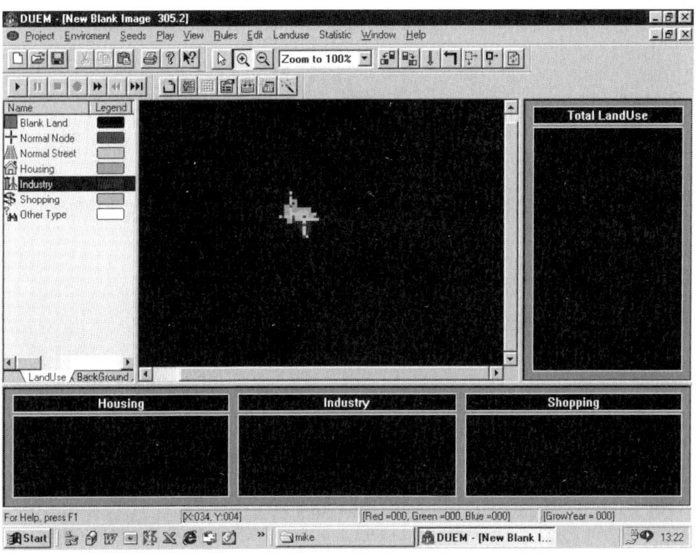

1.2 The city grows through diffusion based on development rules with the growth being capacitated by the size of the space.

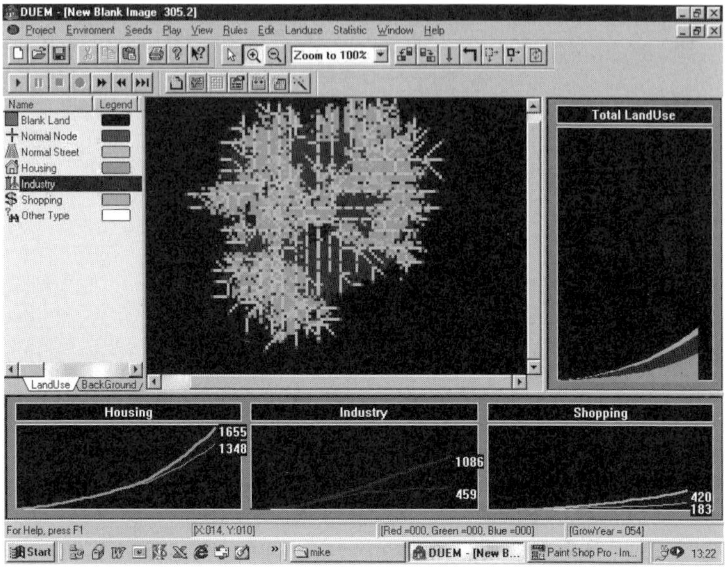

1.3 The city eventually fills the space with housing, industry, and services; then oscillating around the capacity as in logistic growth.

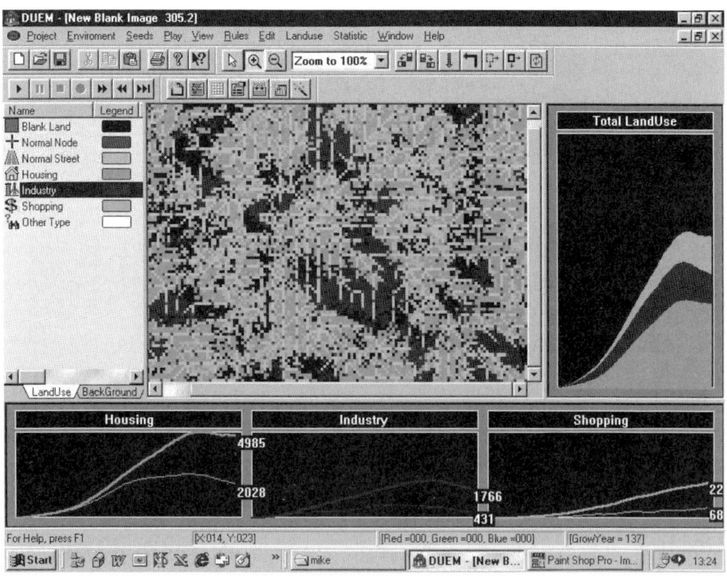

Figure 2. A typical sprawl simulation for the Ann Arbor, Michigan, region.

2.1 Development in Ann Arbor, Michigan, in 1985.

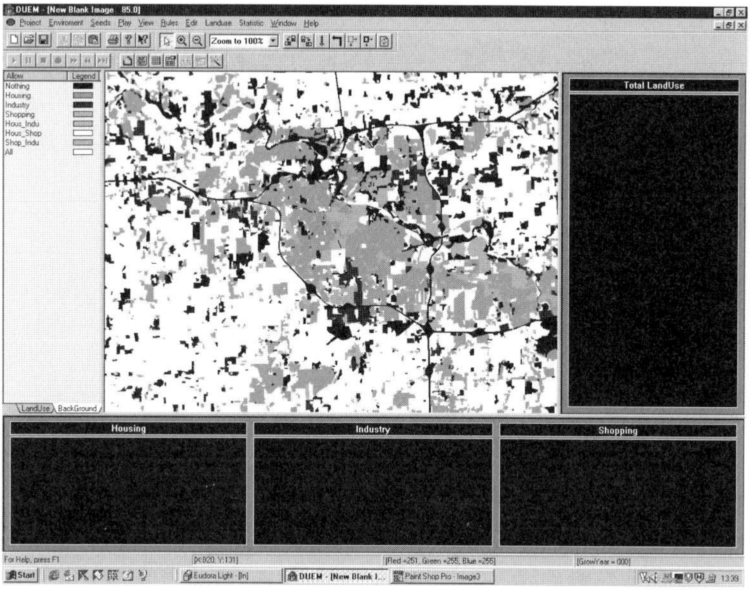

2.2 The seeds for development between 1990 and 1995 based on the change in development from 1985 to 1990.

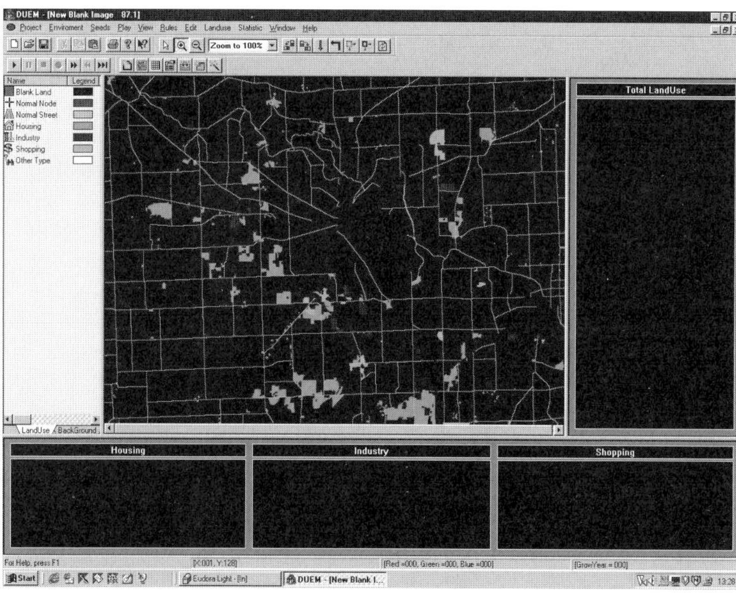

2.3 One growth scenario from 1990 to 1995 illustrating how difficult it is to tune the model to generate realistic morphologies.

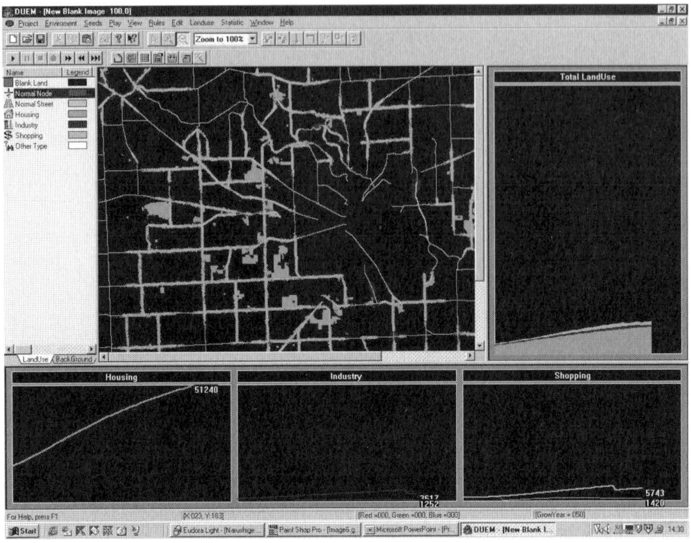

Representation: Models as Data and the GIS Revolution

The problems that led to the demise of the first generation of urban models produced a very different digital cutting edge to urban planning in the 1980s and 1990s. Our inability to deal with large data systems and the unwieldy nature of computer technology in the 1960s was changing as the first generation of urban models became operational, and by the late 1970s it was clear that computer power would never again be a problem in the development of large-scale models. Moore's Law, now enshrined in the history of miniaturization, suggests that computer power (and memory) roughly doubles every 18 months, and by the 1980s, graphics and text processing were becoming the predominant applications. In fact, during the 1980s when many mainframe and workstation software applications were being translated into proprietary form and ported to microcomputers, urban models were conspicuous by their absence. Momentum in the field was at such an ebb that there was little activity in such development. By the time the field woke up again in the mid-1990s, there were few if any applications that utilized the then current power of microcomputing in making generic applications widely available. It might be argued that the market for urban models was too small for such development, but those who were still in the field did not attempt to open their software in this way, and they failed to utilize new graphical user interfaces and the range of map-based software that became available. Only now are such efforts under way.

In terms of the development of computer methods for planners, what in fact happened was the development of more modest tools based on techniques nearer to the data. In many ways, the development of GIS was an obvious step in moving to a digital world, as the processing of maps and storage of their data has very wide applicability, much large than planning per se. GIS originated from a synthesis of three related areas—spatial data base technologies that involved developments in how to represent points, lines, polygons, raster structures, and so on; computer cartography where the concern was the automatic production and generalization of maps across many scales; and ideas in representing layers in the landscape, from landscape architecture and planning where the notion of associating many data layers together had been a central concern in the development of that field. In fact early line plotting of maps as in SYMAP (SymboldMapping) were based on such associations. By the late 1980s, packages were being marketed which had obvious uses in urban planning, if only for automated land-use mapping. But once desktop GIS began in earnest in the early 1990s, coinciding with the Windows operating system, most municipalities began to employ such tools.

The functionality within GIS for modeling and related kinds of urban analysis is still quite limited. Apart from quite elaborate methods for overlaying different maps of data layers and for operating numerically on the process of overlay, most functions need to be added through specialty plug-ins. However, the data layer approach is important in that there are many areas of the planning process, particularly those dealing with growth and location, that require such overlay capabilities. Most of the plug-ins to date involve accessibility and network calculations, spatial interpolation and viewing functions, and more recently extensions to three dimensions, which we will illustrate below. As yet there are few plug-ins for the kinds of tools that were presented a decade or more ago by Brail (1987) in *Microcomputers in Urban Planning and Management,* in which the software used was non-graphical, based on spreadsheets. In fact, the kinds of software that we illustrated in Figures 1 and 2 is still rarely available in that GIS vendors, although intent on adding functionality for planning where there is a market, are less inclined to develop mathematical modeling applications that they still see as somewhat esoteric and controversial in their use.

It is important to note the range of tools that GIS and related software offer planners across their range of tasks. Clearly part of GIS are issues involving the organization of a planning municipality's data bases in more general terms from routine planning applications and permits to the archiving of map data that might be used for public as well as professional services. In terms of the professional planning process, GIS has many uses at different scales from the regional to the local urban design scale and perhaps even down to the building level. Much of this usage depends upon attitudes, for use of digital technologies requires users to think of their problem in abstracted terms, to think of their system as a model and of their professional activity as the use of that model in a problem-solving process where outcomes are then deliberated upon. Moreover, new kinds of data at the fine scale can only be accessed and unlocked using GIS, thus requiring the users to be expert in the digital manipulation of planning data. In Figure 3, we show

Figure 3. Models of the data: Defining a central business district through smoothing and aggregating diverse data layers which define relevant urban activities in Bristol.

3.1 a) Locations of retail activity at unit postcode level in central Bristol, from ArcView.

3.2 b) Interpolated contours of retail activity using the Kernel Density Estimator.

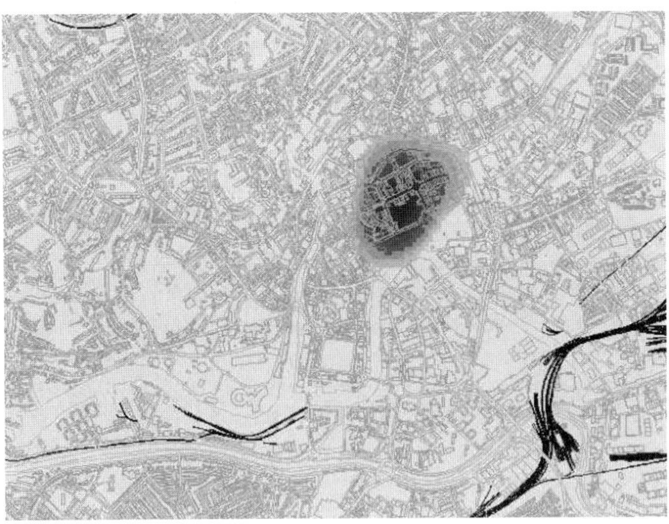

3.3 c) The composite weighted surface compiled by aggregating 42 different services as in b.

3.4 d) The town center boundary as defined by an expert panel using the GIS visualizations.

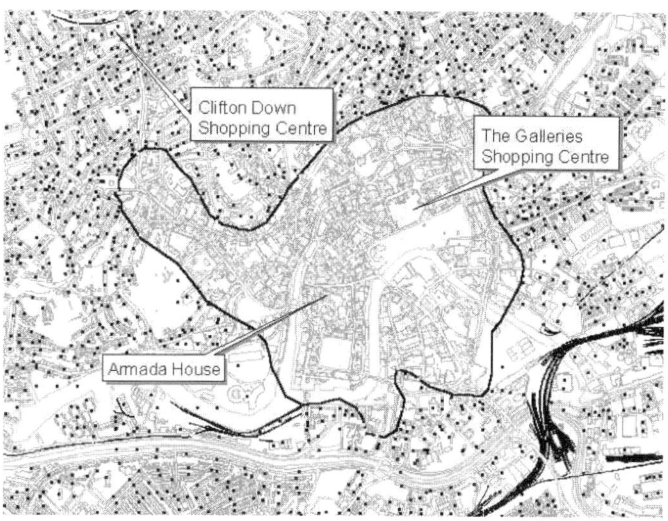

how data associated with many different types of activity defining the Central Business District of the English town of Bristol can be displayed, smoothed, and then aggregated into layers that are used in defining boundaries. This requires considerable knowledge of statistical operations for smoothing that are embodied in the functionality of many contemporary desktop GIS applications, but it also requires professionals who are able to exploit this kind of usage. In short, GIS provides a toolbox which can be adapted to various kinds of applications, only limited by the imagination of the user.

One of the reasons why urban designers have been so slow in adopting such technologies is because the subject matter of their design does not embrace the hard data that is available at the most local level. Moreover, they are unaware of how easy it is to embed multimedia into GIS (Batty et al. 1999). In Figure 4, we show how various kinds of media can be added into desktop GIS for the center of the English town of Wolverhampton. The use of these data, displays, and technologies implies that we need to think of the problem in more abstract terms. GIS of course can be used in much more routine contexts, such as in the control of development but its real power is not so much in providing a purpose-built set of functions for a typical planning task but in providing a toolbox that can be linked to other software in ingenious ways. In this, the prospect of building mathematical models and linking these to the data models provided by GIS, which in turn can be linked to a range of visual and statistical software, probably marks the most appropriate use of these new digital technologies in planning. Moreover, this bottom-up use of software provides a constraint that increases relevance, applicability, and also feasibility in that decisions as to how to combine these low-level tools and functions are left to the user who is engaged in the planning problem.

This shift to the low-level models that have largely taken the place of those operational mathematical models fashioned a generation or more ago is not all GIS-based. In fact, although GIS now represents the most effective focus of this kind of basic tool, other software has and continues to be used extensively within urban planning. Spreadsheets represent the most widely employed generic tool for mathematical modeling as reflected in the variety of applications collected by Klosterman, Brail, and Bossard (1993). Recently, various plug-ins such as those based on the GIS application MapInfo have been linked to the spreadsheet Microsoft Excel and these kinds of software continue to expand to embrace yet more statistical and related tools. It is now possible to develop extremely elaborate models within a spreadsheet and to link these to a variety of other graphics media. Many possibilities abound and the fact that there are so few applications probably reflects the fact that the range of possible applications is so great in comparison to the numbers of researchers and users. Few have the time and ability to fashion such extensions, and there are enormous opportunities for the development of new software and applications in the domain of urban planning. Other software is also worthy of note, such as the development of graphical model–building software from Systems Dynamics–like programming such as STELLA to full-fledged programming environments such as that contained within Mathematica.

Figure 4. GIS for urban design: Articulating the local design factors digitally in the town center of Wolverhampton, U.K.

4.1 a) Digital data at 1:1250 for central Wolverhampton.

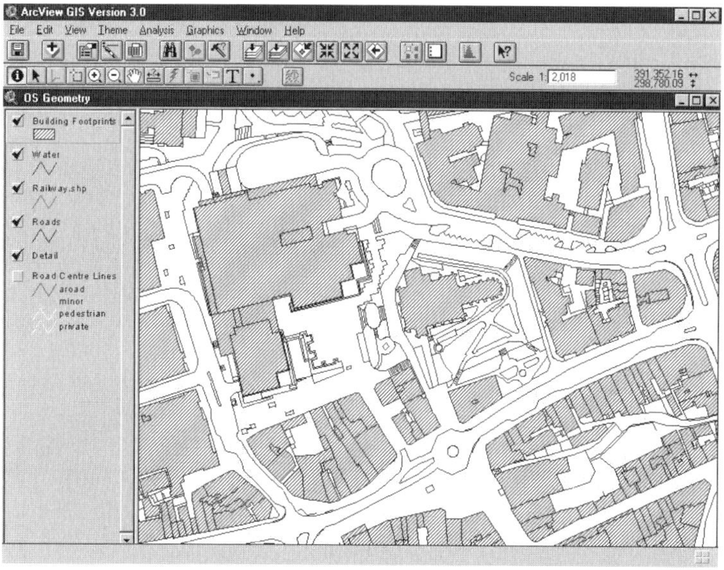

4.2 b) Digital aerial photography of the northwest corner of the center.

4.3 c) Land uses in the center.

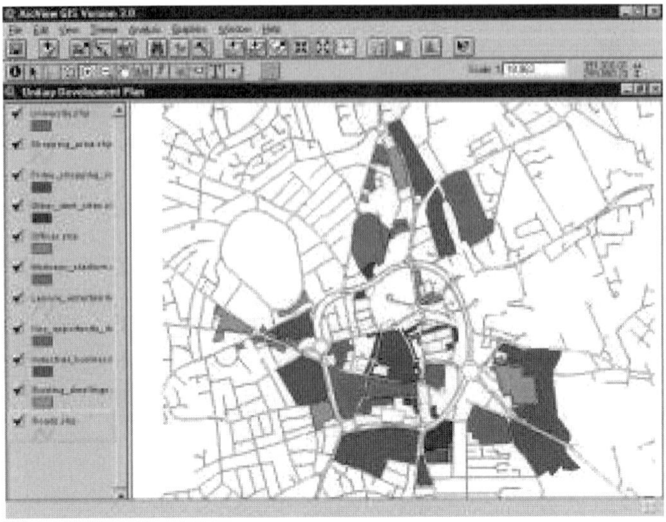

4.4 d) Digital video embedded within the GIS.

Before we turn full circle and show how this new representational point of view is beginning to collapse back into the use of icons, it is worth noting that many of these programming and data environments that imply low-level modeling are in turn changing. Increasingly, every piece of software is becoming open to every other and it is becoming possible to develop generic models in many different media. Moreover, the way models and data are now being communicated and accessed is increasingly important—not only on the desktop but over the net and in various immersive contexts—and this is beginning to change not only what is simulated and what is communicated but also who is involved in model design and use. We will return to these issues in the penultimate section, where we examine how computers are opening up the planning system to wider participation but before this, we must review where we stand on new forms of representation.

Moving Models Back to Icons: Representations, Images, Digital Toys

The idea that computers could be used for graphics goes back to the very beginning of the computer era when the pioneers used oscilloscopes for monitoring the workings of the first machines and realized that the scopes could be programmed to produce pictures of various kinds. Indeed the very first arcade game, Pong, commercialized in the early 1970s by Atari, was one of the first graphics ever illustrated from a digital computer being shown on U.S. television in the early 1950s. Although it took the micro-revolution to raise graphics to the level where it has become the dominant digital technology, the first attempts at putting the traditional iconic model of the city into a computer go back to earliest days of computer-aided drafting (CAD). The wire frame fly-through of buildings in downtown Chicago by Skidmore, Owings, and Merrill in the early 1980s showed what was possible, but widespread use of these kinds of 3-D technologies did not occur until quite recently.

CAD began almost as soon as the personal computer invaded the market. For example, many 3-D architectural and city models have been developed in AutoCAD and with the growth of the web various internet equivalents, such as VRML, are being used for display and navigation. However such models are relatively superficial and their usage has been confined to presentational purposes. Although such CAD invokes the idea of design, such design is always offline. In fact, contemporary CAD models of cities are rather limited in that all one can do is use them for visualizing changes to the urban geometry. The geometric structures that they form rarely have any content—data—other than that associated with building rendering, and hence their use in anything other than design review is limited.

The thrust that is returning planning and urban design to these digital icons is not from CAD but from GIS. Three-dimensional visualizations represent another way of viewing spatial data and in many senses these complement the 2-D map view. Many GIS packages now have extensions that enable 2-D map data to be extruded into 3-D where the 3-D content is some spatial attribute. Spatial patterns and errors in data can be more easily visualized in 3-D, while the ability to pan and zoom enables detail to be examined in a way that is not possible in a map view. In fact, 3-D views are very effective ways of looking at multiple patterns in the data. For example, the density of any activity is often taken to mark the third dimension and if this is used, then a further variable can be coded into the scene. If several such scenes are displayed together, some packages offer ingenious hot-linking that enables users to explore the data by brushing and examining correlations and other associations between different scenes. Figure 5 shows population density in the London Borough of Westminster, which is illustrated as a 2-D map and 3-D block model using the plug-in 3D Analyst, which is part of the desktop GIS ArcView.

Recently we undertook a worldwide review of 3-D digital models for the Corporation of London (Batty et al. 2001) and concluded that CAD models no longer represented the cutting edge of such technologies. Users now require iconic models in digital form as a means to accessing data, rather than for generating aesthetic impacts or providing visual form to their designs. Upwards of 80 serious applications now exist worldwide with GIS and various spatial database technologies representing the predominant means for such model construction. Realism in terms of rendering, though important, is also being automated through various photorealistic imagery while height data is being sourced through remote sensing of various sorts. For example, Figure 6 shows a 3-D visualization of the area around St. Paul's Cathedral in the City of London. The building blocks have been taken from crude LIDAR data, the cathedral itself being constructed from the several thousand points generated by the laser-based technology. This is the state-of-the-art and very shortly such models will be generated on the fly as satellite data at the requisite level of resolution becomes available on a daily basis.

Figure 5. From 2-D to 3-D: Population density in Westminster (London) generated as block model within desktop GIS.

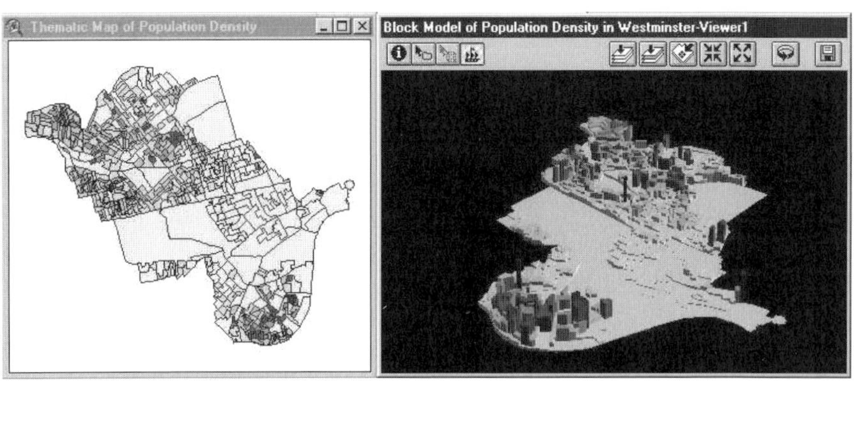

Figure 6. A GIS 3-D block model of the St. Paul's District in the City of London with height data from LIDAR data.

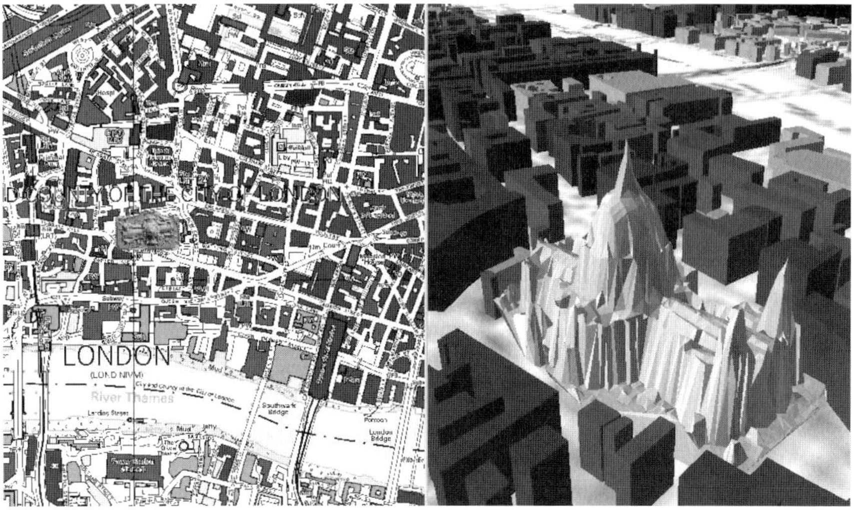

One of this paper's main themes involves the way the idea of the computer model of the city has changed during the last 50 years, and our implicit argument that we have come full circle is clearly being borne out by the flurry of 3-D digital block models of cities currently being constructed. Yet this circularity is never what it seems. This is no mere return to a world of the architect's models where the focus is purely on visual evaluation. As we shall see in our concluding section, access to these new digital icons is very different from the icons of the pre-computer age. The visual simplicity of their form belies their true purpose, which is much more likely to be as an interface to a complex and interrelated data base where 3-D visualization is but one way of accessing the data and managing this complexity.

However, there is another set of applications which is likely to convolute these applications even further. We have already implied that new functionality is being introduced into GIS that is extending their use from representation to simulation. The same is occurring in 3-D. Already there are traffic models where the representation enables actual car movements to be simulated mathematically and visualized in 3-D while there are experiments to show how various kinds of model process can be embedded within 3-D virtual environments of the kind visualized above. This is the point at which the models themselves must match the level of representation and it implies that the kinds of models that are suitable are those that are less abstract, more micro in scale and form, and deal with routine dynamics processes that might be captured in dynamic realizations of the 2-D map or 3-D geometry in question. For example, work at the Environmental Simulation Center in New York City for the town of Scutney in New England has developed a visualization of an agent-based model of community development, tying its mathematics back to data as well as to 3-D icons. We ourselves designed various models of the movement of agents in shopping centers and in galleries and have visualized such movements in 3-D. In our models of movement in the Tate Gallery (Figure 7) we visualized agents moving in both 2-D and 3-D, examining different digital visualization and simulations of the same system (Batty et al. 1998). In the next decade we are likely to see many new fusions and extensions of the mathematical, representational, and data models that we have through different kinds of digital media, many not yet invented.

Moving Models Back to Users: Participating in Digital Design

In public policy particularly, there has always been concern as to the relevance and remoteness of the science with respect to the kinds of advice that such science provides for issues that affect a wider public. However, attempts at developing mathematical models of urban systems in the wider context have been limited. Some attempts at providing environments in which systems dynamics models might be developed jointly with decision makers have been tried but in general, such efforts have been labored. However, the new focus on visualization that GIS and VR are providing and the new concern to open up data and symbolic modeling using visual interfaces is at last providing a context in which interests other than those close to the science are able to make an impact. The notion too of providing more than one approach to the same problem is also intimately associated with this process of participation. The high degree of uncertainty in human systems is such that more than one model is usually required to provide a balanced perspective on policy making. The process of counter modeling first proposed by Greenberger, Crenson, and Crissey (1976) more than 20 years ago is now much more of a reality in a digital world where different perspectives on any problem can be quickly fashioned from the use of different computer-based tools.

There are as yet few forums in which a range of users can build models, but the World Wide Web provides enormous potential for different users to generate their own interpretations of data and, in principle, build their own models, either individually or in the presence of others. As we have noted, desktop software of various kinds is being ported to the web. Internet GIS and CAD are gradually being established and this is delivering visualizations and map data that can be viewed and acted upon by a wide range of interests. As an example of what is possible, many of the visual techniques that we have presented in this paper are being used in an inner-city context for the development of awareness about problems of regeneration and deprivation in the London Borough of Hackney. The Hackney Building Exploratory is a project built around conventional approaches to raising environmental awareness through non-digital maps and designs and related products in the local community. This is now being supported by a complementary digital initiative aimed at public participation using digital data and icons to

Figure 7. Modeling agents by simulating movement in 2-D with visualization in 3-D.

7.1 a) Walkers in the Tate (left) August 16, 1995, and simulations of the same using an agent-based model (right).

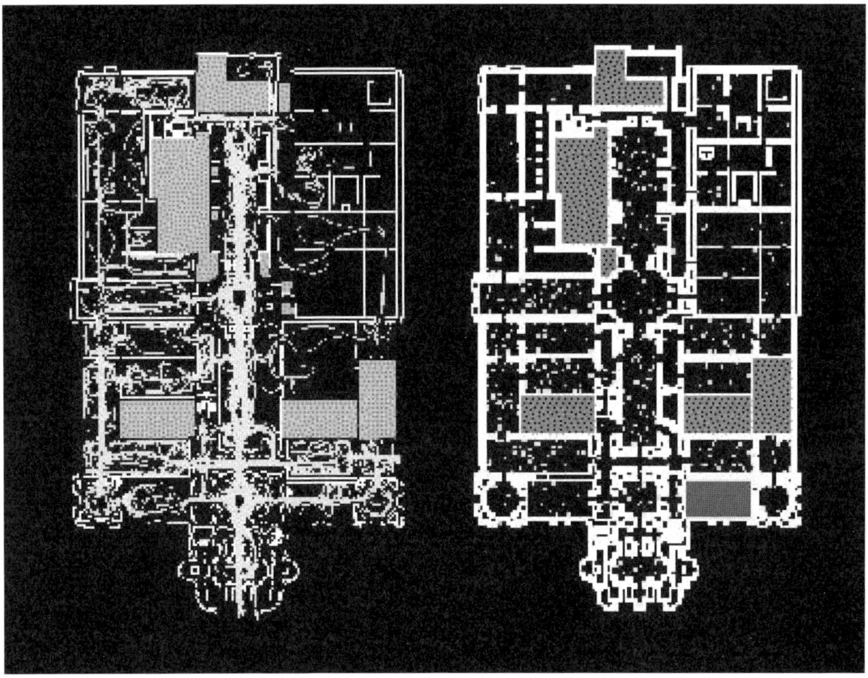

7.2 and 7.3 b) Predicted density of traffic in the Tate from the agent model (left) and the visualization in 3-D CAD model (below).

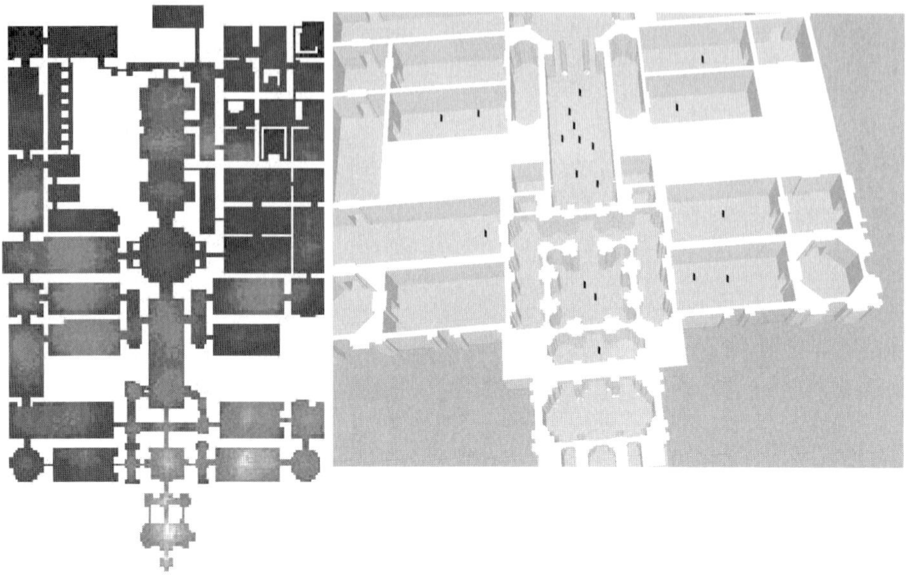

7.4 c) Visualization of agents walking in a realization of the Tate as a virtual world.

support the process. In Figure 8, we show some images from this project which reveals how VR, CAD, and GIS can be used to unlock data and to deliver this to users across the web. This data varies in scale but the intent is to educate and communicate issues of concern in the local community. We have not yet sought to provide users with model-building capability, but in the way they are encouraged to put data together there is an implication that models are being built that reflect the interests and beliefs of those who are being affected by local planning issues.

In Figure 7, we illustrated how models of movement could be linked to virtual worlds and an obvious extension of such worlds involves ways in which users might themselves enter such environments alongside their simulated selves. This sounds like science fiction—the juxtapositions of remote but real individuals with simulated agents and their interactions—but these kinds of environments are now possible and open up a variety of ways in which simulations might be tested and modified in ways that have hitherto not been possible. There are many possibilities emerging from the digital world being constructed, and all promise to change the way we involve ourselves in public affairs and urban policy. The only limits to such possibilities are in terms of what we might imagine and this is a far cry from the earliest use of computers in planning and public policy which saw their use in purely passive, top-down contexts. In the next 20 years, there promises to be a fusion of ideas which will broaden the concept of models in planning even further.

Figure 8. Bringing digital data and models to users: The Hackney Building Exploratory—interactive.

8.1 a) The homepage.

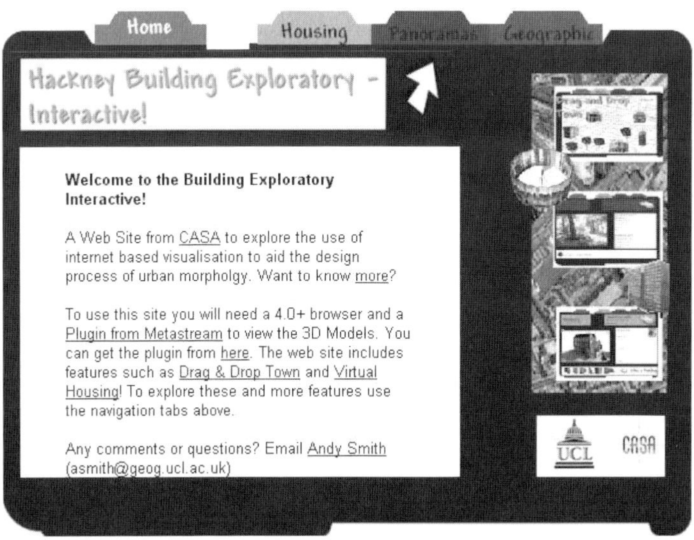

8.2 b) A house type in wire-frame mode.

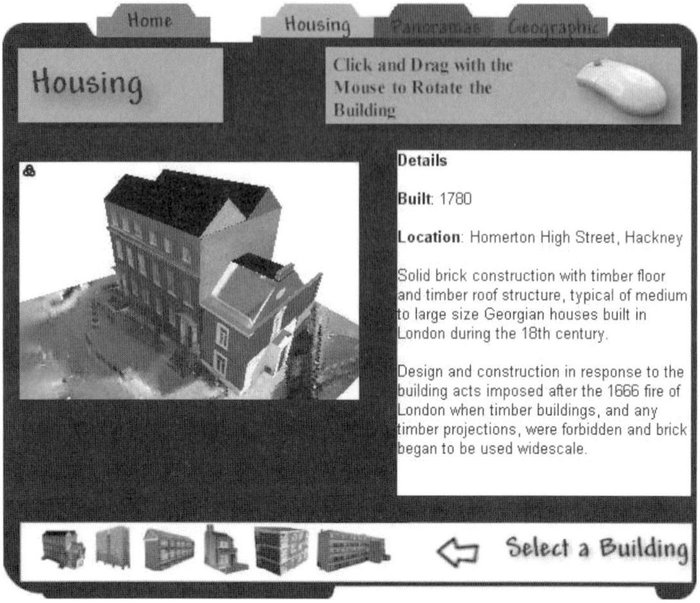

8.3 c) A street panorama.

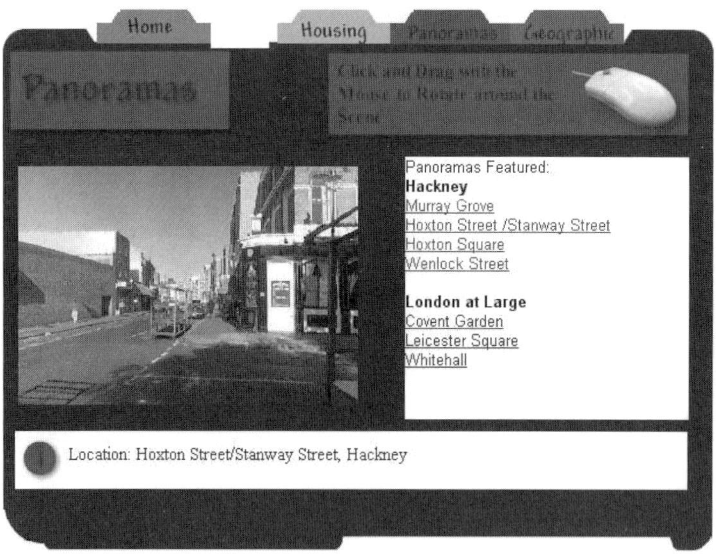

8.4 d) A block model of social deprivation.

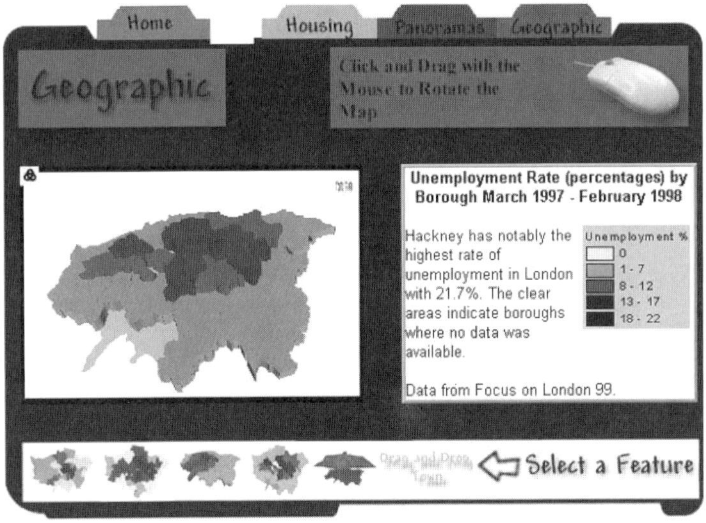

8.5 e) The original drag and drop town canvas.

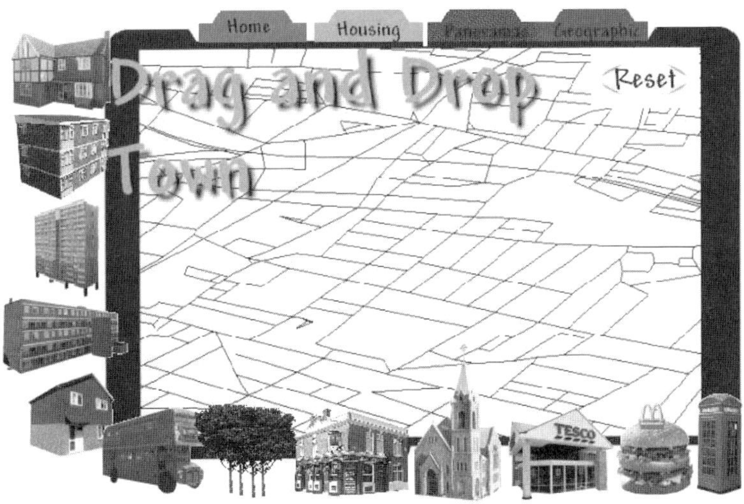

8.6 f) A user design.

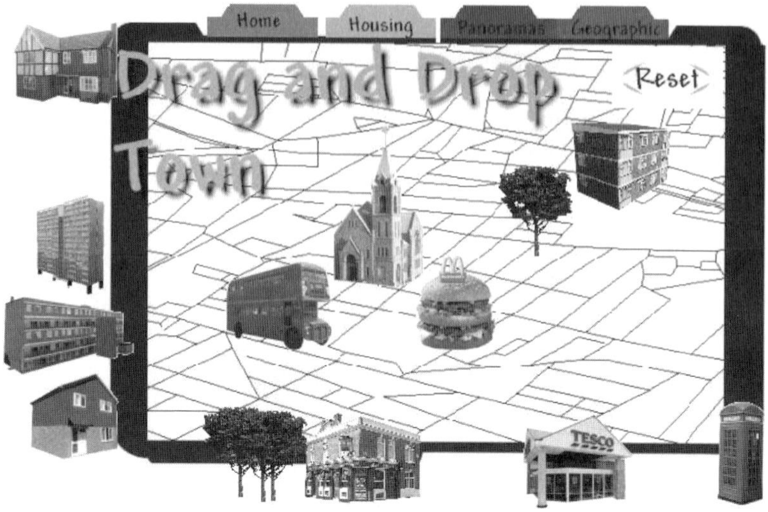

Conclusions: Models and Modeling for Cities of the Twenty-first Century

The critique of models that we have developed in this paper is no more trenchant that those that have been leveled at every area of the human sciences. The effort to come to grips with social complexity in terms of its manifestation as urban phenomena, and the ideology that supposes that the human condition can be bettered by interfering in the process of urban development, is fraught with difficulty as the experience of the last 50 years have shown. We still do not have good theory about how cities function and although there are signs that researchers are becoming more aware of their intrinsic complexity, the models that have been developed are as arbitrary as those which form the conventional wisdom of contemporary economics. Moreover, human systems are ever changing as their populations innovate and develop new sets of preferences and tastes, thus laying waste to theory that once seemed applicable. The current concern for theories that make sense of complexity and change on all scales is indicative of the fact that it is virtually impossible to make meaningful predictions for such systems or at least predictions that would form the basis of medium- or long-term policy making.

New theories of cities are urgently required but these need to be developed in much more intensive data environments than anything that has yet been used. In fact, urban studies must move to a new plane in terms of the way it handles data and develops theory, more akin to the way large-scale data systems in the physical sciences are being handled. To this end, many of the representational techniques presented in the latter part of this paper are essential in unlocking complex data sets and developing methods to search for new data patterns that might provide the basis for new notions of causality and association. As ever, the world cannot wait for new theories which, in any case, are dependent on the very world that they seek to explain. Thus what is required is a strategy of theory building, modeling, and prediction in planning that seeks to arrange techniques and models in parallel and in hierarchy, with different ideas tailored to different scales, problems, sectors, and processes. Counter-modeling and modeling in parallel are now essential and the data-rich environment which we have sketched here has all the potential to provide the basis for this new understanding. This we suggest will mark the next generation of urban research and policy making.

Acknowledgements

This research has been partly financed by the ESRC NEXSUS Project: Network for Complexity and Sustainability (L326253048).

References

Allen, P. M. 1997. *Cities and regions as self-organizing systems: Models of complexity*. Amsterdam, The Netherlands: Gordon and Breach.

Batty, M. 1994. A chronicle of scientific planning: The Anglo-American modeling experience. *Journal of the American Planning Association* 60:7–16.

———. 1979. Progress, success and failure in urban modelling. *Environment and Planning A* 11:863–878.

———. 1976. *Urban modelling: Algorithms, calibrations, predictions*. Cambridge, U.K.: Cambridge University Press.

———. 1972. Recent developments in land use modelling: A review of British research. *Urban Studies* 9:151–177.

Batty, M., D. Chapman, S. Evans, M. Haklay, S. Kueppers, N. Shiode, A. Smith, and P. Torrens. 2001. Visualizing the city: Communicating urban design to planners and decision-makers. In *Planning support systems: Integrating Geographic Information Systems, models and visualization tools*, edited by R. Brail and R. Klosterman. New Brunswick, N.J.: Rutgers University Press and Redlands, Calif.: ESRI Press.

Batty, M., R. Conroy, B. Hillier, B. Jiang, J. Desyllas, C. Mottram, A. Penn, A. Smith, and A. Turner. 1998. *The virtual tate*. Working paper 5. London: Centre for Advanced Spatial Analysis, University College London. Web document available at http://www.casa.ucl.ac.uk/tate.pdf.

Batty, M., M. Dodge, B. Jiang, and A. Smith. 1999. Geographical Information Systems and urban design. In *Geographical information and planning*, edited by J. Stillwell, S. Geertman, and S. Openshaw. Heidelberg, Germany: Springer.

Batty, M., and P. Longley. 1994. *Fractal cities: A geometry of form and function.* San Diego, Calif.: Academic Press.

Batty, M., Y. Xie, Z. and Sun. 1999. Modeling urban dynamics through GIS-based cellular automata. *Computers, Environments and Urban Systems* 23:205–233.

Ben Akiva, M., and S. Lerman. 1985. *Discrete choice analysis: Theory and application to travel demand.* Cambridge, Mass.: MIT Press.

Brail, R. K. 1987. *Microcomputers in urban planning and management.* New Brunswick, N.J.: Rutgers University Press.

Brewer, G. D. 1973. *Politicians, bureaucrats and the consultant: A critique of urban problem-solving.* New York: Basic Books.

Chapin, F. S., and S. F. Weiss. 1968. A probabilistic model for residential growth. *Transportation Research* 2:375–390.

De La Barra, T. 1989. *Integrated land use and transport modelling.* Cambridge, U.K.: Cambridge University Press.

Dendrinos, D. 1992. *The dynamics of cities.* London: Routledge.

Echenique, M. H. 1994. Urban and regional studies at the Martin Centre: Its origins, its present, its future. *Environment and Planning B* 21:517–534.

———. 1972. Models: A discussion. In *Urban space and structures*, edited by L. Martin and L. March. Cambridge, U.K.: Cambridge University Press.

Environmental Simulation Center. 2000. Web document available at: http://www.simcenter.org/Projects/CommunityViz/communityviz.html. Viewed ADD DATE.

Forrester, J. W. 1969. *Urban dynamics.* Cambridge, Mass.: MIT Press.

Greenberger, M., M. A. Crenson, and B. L. Crissey. 1976. *Models in the policy process: Public decision-making in the computer era.* New York: Russell Sage Foundation.

Harris, B. 1994. The real issues concerning Lee's "Requiem." *Journal of the American Planning Association* 60:31–34.

———. 1968. Quantitative models of urban development: Their role in metropolitan policy-making. In *Issues in urban economics,* edited by H. S. Perloff and L. Wingo. Baltimore, Md.: The Johns Hopkins University Press.

Hill, D. M. 1965. A growth allocation model for the Boston region. *Journal of the American Institute of Planners* 31:111–120.

Klosterman, R. E. 1994. Large-scale urban models: Retrospect and prospect. *Journal of the American Planning Association* 60:3–6.

Klosterman, R. E., R. K. Brail, and E. G. Bossard, eds. 1993. *Spreadsheet models for urban and regional analysis.* New Brunswick, N.J.: Rutgers University Press.

Landis, J. D. 1994. The California urban futures model: A new generation of metropolitan simulation models. *Environment and Planning B* 21:399–420.

Lee, D. B. 1994. Retrospective on large-scale urban models. *Journal of the American Planning Association* 60:35–40.

———. 1973. Requiem for large-scale models. *Journal of the American Institute of Planners* 39:163–178.

Lowry, I. S. 1965. A short course in model design. *Journal of the American Institute of Planners* 31:158–166.

———. 1964. *A model of metropolis.* RM-4035-RC. Santa Monica, Calif.: The Rand Corporation.

Massachusetts Institute of Technology, Media Lab. YEAR. Web document available at http://www.media.mit.edu/toys/totweb/index.html.

Portugali, J. 2000. *Self-organization and the city*. Heidelberg, Germany: Springer-Verlag.

Putman, S. H. 1991. *Integrated urban models 2: New research and applications of optimization and dynamics*. London: Pion.

———. 1983. *Integrated urban models: Policy analysis of transportation and land use*. London: Pion.

Schock, S., ed. 2000. *Projecting land use change: A summary of models for assessing the effects of growth and change on land use patterns*. EPA/600/R-00/098. Washington, D.C.: National Exposure Research Laboratory, Office of Research and Development, Environmental Protection Agency.

Simon, H. A. 1991. *Models of my life*. New York: Basic Books.

———. 1982. *Models of bounded rationality*. 2 volumes. Cambridge, Mass.: MIT Press.

———. 1979, 1989. *Models of thought*. 2 volumes. New Haven, Conn.: Yale University Press.

———. 1977. *Models of discovery and other topics in the methods of science*. Dordrecht, Holland: D. Reidel.

———. 1957. *Models of man*. New York: John Wiley.

Waddell, P. 2000. A behavioral simulation model for metropolitan policy analysis and planning: Residential and housing components of UrbanSim. *Environment and Planning B* 27:247–263.

Wegener, M. 1994. Operational urban models: State of the art. *Journal of the American Planning Association* 60:17–30.

Wilson, A. G. 1998. Land use–transport models: Past and future. *Journal of Transport Economics and Policy* 32:3–26.

———. 1981. *Catastrophe theory and bifurcation: Application to urban and regional systems*. Berkeley: University of California Press.

———. 1974. *Urban and regional models in geography and planning*. Chichester, U.K.: John Wiley.

Wilson, A. G., J. D. Coelho, S. M. Macgill, and H. C. W. L. Williams. 1981. *Optimization in locational and transport analysis*. Chichester, U.K.: John Wiley.

Wittgenstein, L. 1921, 1961. *Tractatus logico-philosophicus*. London: Routledge and Kegan Paul.

Integrating Knowledge about Land Use and the Environment Through the Use of Multiple Models

Lewis D. Hopkins

Introduction

My fundamental premise is that reaching focused disagreements and then explaining them is fundamental to crafting research agendas and to using urban and environmental models. We should seek neither one integrated research agenda nor one integrated model. We should, instead, create computer modeling environments in which disagreements can emerge and be used to advance knowledge and be used to solve problems. The first use is a way to organize research about land use, environment, and the use of models. The second use is a way to decide what to do in specific situations—to apply the knowledge and models. We should view models and computing tools as additional participants in conversations—interactions about discovering and using knowledge.

Reaching and using disagreement requires:

- Frameworks within which different ideas or different models can be considered
- Mechanisms for translating from one set of ideas to another or from one model to another
- Interfaces that enable consideration of multiple ideas or models

Luckily, these requirements are approximately the same for both enabling knowledge advancement and enabling problem solving: create a comparative framework that puts different things in the same terms to discover how they differ, what they can tell you, and how each may help to point out, compensate, or correct the others.

Verma (1998) argues that we should focus on similarities and connections rather than difference. He argues in particular for the use of analogies—identifying different situations focused on similar purposes. Is the idea of reaching disagreement a different approach? Finding an analogy is equivalent to finding a different model that claims to accomplish the same purpose. An analogy may lead to a direct solution if the situations are really exactly the same, but it may also lead to recognition of how two situations are similar in some ways and different in others. Verma argues against a focus on distinctions that result in only one analytical perspective being applied to a problem. Although my focus on differences might appear to be the opposite of his focus on similarity, I think these two foci have much in common. Both imply frameworks in which we can organize different claims or different models about the same phenomena or issues. We should not seek one analogy, but several different analogies. We need to be able to figure out how the situations or models differ, not to focus on one model or analogy alone, but to take advantage of what these differences tell us. This approach involves using several of the models, not choosing one.

Comparative Frameworks: Examples from Research

The following examples illustrate how this approach frames research. They also suggest how planning support systems can incorporate several models as replicates and complements.

In Hopkins (1977b) I presented a framework in which eight ways of estimating the suitability of land for a particular use could be estimated. Identifying the similarities and differences among these methods in a common framework made clear their particular characteristics and their advantages and disadvantages. Using more than one method is one way to overcome the limitations of each in practice.

In a series of papers on land allocation (Hopkins 1977a, 1979; Hopkins and Los 1979) we used a data structure that allowed several different optimization and simulation models to be used in combination. Some of the models were used iteratively to generate land use patterns likely to perform well on modeled criteria, which was useful in exploring questions about good land use patterns. These models included quadratic assignment, central facility, and linear programming models. Spatial interaction and spatial diffusion models were used to generate data and assess resulting land use allocations. Using models in this way required a common framework for data. It required translations for the inputs of one model to the inputs of another so that the models could be compared as alternative representations of the same situation. It required translations from the outputs of one model to the inputs of another so that the models could be used iteratively to converge on land use patterns that considered several different issues represented in different models. And, it required translations from the output of one model to the outputs of another in order to compare the implications of two different models for the same situation. Each model type had advantages in addressing aspects of the situations, but no model was able to address all aspects of interest. It is more useful to generate alternative land-use patterns from a set of different models and interpret the implications than to pretend to capture all aspects in one model.

We also developed techniques explicitly designed to generate different solutions (Hopkins et al. 1982; Brill et al. 1982; Brill et al. 1990). These tools relied on a single model. They had the advantage, however, of treating the model as only an approximation of the problem. These models generated and presented to decision makers alternatives that were good with respect to modeled criteria and very different from each other in spatial pattern. Planners and decision makers can then use other models to analyze these patterns with respect to other issues not addressed by the model. A remaining research question is: How can computing capabilities be used to enhance our ability to generate alternative models of the same situation? We return to this possibility in the next section with examples of alternative models that might be generated for environmental problems.

Another set of papers focused on multiattribute decision making, an extension of the earlier work on suitability analysis. Again, we created a framework in which a wide range of techniques could be compared, translated one to another, and used together to point out disagreements in their implications (Lai and Hopkins 1989, 1995; Lee and Hopkins 1995). This work included experiments made possible by the common framework and translations because subjects could use different techniques iteratively. Thus, we could compare judgments made by individuals using different methods and control for order effects. Our results suggest that individuals cannot make judgments consistent with the logic of the methods they employ and that the method used affects the judgments made. One response is to provide decision support tools that encourage use of more than one multiattribute method and help to think through why the methods yield different preferences and choices.

Westervelt (2001) describes a modeling environment that supports linking models developed by many different teams working separately and in different disciplines. The mechanisms for linking models include spatial, functional (process), and temporal frames. The premise is that there are many disparate models available to address aspects of a watershed system. We can leverage these modeling assets by creating computing environments in which to use them. A first level is to provide a common input and output frame (a data structure) that can encompass different models. Even more powerful will be a frame in which different models can interact as processes, not merely as inputs and outputs. This level of interaction will require a common data structure of modeled entities rather than the current focus on a common spatial cross section from GIS. We return to this question in the discussion of a research agenda.

Building on these examples, we next consider how multiple models might be used in the environmental and land use area. Then we draw from these ideas to suggest an agenda for research.

Using Multiple Models

What could a computer modeling environment provide to increase our capabilities to use multiple models? One answer is that it could present us with alternative models by breaking some of the assumptions that constrain our initial modeling. Two examples emerged from conversations with participants in the conference. They illustrate the potential of working with different models rather than focusing on building an integrated model. The first considers a question of forecasting, the second a groundwater problem. In the first example, reversing the sequence of independent and dependent variables might have a significant effect on our thinking.

A computer modeling environment could plausibly compute such a reversal and present it to us, even if we did not ask for it. The second example implies a computer modeling environment with domain specific knowledge, which could be used to present to us an alternative physical model with similar input and output variables but different physical processes. It is worth noting that for either example we could focus on organizing our human collaborators to play these roles rather than designing computer collaborators to do so. We should do both.

In the desert Southwest, we face a problem: A growing population requires water. A typical formulation of this problem is to focus first on predicting future population and future water use per capita. Then we ask how such a supply of water could be supplied physically, how this water might be priced, or how the population increase might be controlled. In this formulation we are likely to focus on getting better data about variables that affect population growth. The better these independent variables, the better our population forecast. Once we have an accurate population forecast, we then consider the implications for water supply. That is, we work in sequence from a set of independent variables to predict population as a dependent variable, then population becomes an independent variable to which we add others to predict demand for quantity of water, or perhaps demand for quantities with respect to price.

A computer modeling environment could tell us to reverse this perspective. Start by trying to estimate water supply under various scenarios, then predict the population that could be supported—attracted or constrained—by water supply, either as a physical quantity or a supply curve with respect to price. In this frame, we are more likely to focus on improving a set of independent variables that would help predict water supply as a dependent variable, which would then become an independent variable, along with others, from which we predict population. We will learn more about how the system works and make better decisions if we model this situation both ways, and ideally a few more as well. One formulation of the problem focuses effort on one set of questions and another focuses it on another. We tend to take one point of view, especially if we are trying to build an integrated model that considers everything. It is so hard to build such integrated models that we cannot also build counter models to discover what blinders our integrated model created.

Assume that you are investigating, for purposes of knowledge generation or decision making, the movement of underground water and thus the potential movement of contaminants that affect its use as drinking water. You have several observation wells near potential contamination sources and several others near potential damage destinations. Well cores in both these locations are mostly sandstone. You devise a model based on the physical processes of groundwater movement in sandstone. You fit this model quite successfully to the available data from source area cores and destination area cores. You then use the model to predict the effects of changes in withdrawal rates, changes in rainfall due to global warming, or changes in contamination sources. The success of the model increases your confidence in what your are doing and narrows your focus to the model rather than the world of which it is an abstraction.

A computer environment could suggest to you the possibility of a movement model based on fracture zones or some other physical process, which might occur in the zone between your well cores. Or, the computer modeling environment could enable you to fit some other researchers' groundwater flow models for sandstone. In either case, these other models would have to work from the same input variables and yield the same output variables or rely on translations of variables across models. It is then possible to discover and consider differences in the results from the different models. There are likely to be differences in the results because the models are based on different assumptions about physical processes or different implementations of similar processes. Do these different models matter? How can differences be explained? For what questions would more knowledge be most useful in resolving or coping with these differences? Again, the important point is that we are likely to keep going on our merry way toward a narrow, overconfident view, especially if we are trying to build one integrated model. We need a work environment of colleagues and computer models that will give us well-timed kicks in the thought processes.

Choosing Sets of Models

For any one question or problem, we have many urban models or environmental models from which to choose. Choosing models to link into more integrated models or to use as complements for different aspects of a system implies one set of requirements. As suggested by Westervelt (2001), these include comparable or translatable time, space, and input and output variables. In choosing sets of models as mutual counter-models—models likely to provide contrasting information that can lead to more understanding and avoid narrow views—the requirements are different. We cannot just choose models on the basis of compatibility of inputs and outputs or of complementarity of subject matter. We cannot say beforehand exactly what differences among models will matter. A good approach is to select models that "mean" different things. What do urban development models mean? What are each model's underlying structures and assumptions that frame its view of how the world works?

Many models of urban development have been devised over the last 40 years and there are several good reviews from various perspectives (Anas et al. 1998; Foot 1981). Wegener (1994) reviewed 12 urban models, considering the subsystems included in the model, the theory on which the model is based, and the policies that the model could assess. In general, all the models are intended to estimate patterns of urban development, but they do so in quite distinct ways. Some simulate markets for land, housing, and labor. Others rely on a priori rules for the likelihood of land conversions from one use to another. Some consider preferences of households and firms based on past behavior with respect to prices. Others rely on past probabilities of land conversions. Some seek an equilibrium solution, implying a long enough time period and foresight sufficient to reach such an equilibrium despite the hard-to-reverse, indivisible decisions about urban infrastructure and buildings. Others are dynamic, usually in the form of recursive, discrete time intervals in which actions depend on the results of previous time intervals. Some model trip behavior given locations; others locate housing based on accessibility. Some include employment and transport of goods; others do not. Although all these models try to say something about how cities grow and change, the models have different meanings. Solving for a market equilibrium answers a fundamentally different question than projecting past land conversion probabilities. Modeling employment location and housing location answers a fundamentally different question than modeling housing given employment.

For triangulation purposes, we should choose models with fundamentally different meanings. Then we should figure out how to translate among their data inputs and outputs so as to apply them to the same datasets and policy questions. Each model must be translatable into at least one other model on a connected network of translations. We do not need direct translations from each model to each other model. Models constructed as object-oriented software will be easier to translate into other models. For this reason UrbanSim (Waddell 2000), which is object-oriented, disaggregated, and fairly inclusive of urban subsystems, is likely to be a good reference model. It provides a starting point from which to elaborate a data structure that could translate it into variables for other models and vice versa. Translations will be imperfect, but recognizing and reporting imperfections is one way to discover useful disagreement.

We can thereby create a common framework and a set of translations so that we can ask: What do these models mean relative to each other? How should their results differ and why? Roughly analogous to the comparison of land suitability analysis methods or multiattribute methods, we ought to be able to explain in comparative terms how these models relate to each other. With such a framework we can discover what we don't know—what we are unable to explain about the differing results. For application, we can use the models as triangulators for each other to improve our understanding of the situation to increase the scope of issues considered.

A Research Agenda

One research agenda for land use and environmental modeling can be described in two major components quite distinct from the idea of creating an integrated land use and environmental model: 1) establish questions, problems, and test cases with data for each question or problem; and 2) create a computer modeling environment with sufficient scope for these questions and problems. The first establishes specific instances or games in which to compete. The second establishes a playing field on which to cooperate in order to compete with our own models to discover what our models can do as complements and as replicates. The opportunity of structured competition motivates participation and focuses disagreement on questions or problems on which we believe we can make collaborative progress.

First, we should identify and develop a set of use cases for modeling. These uses of modeling set the scope of the modeling environment needed to cope with a range of models, questions, and problems. This scope must have limits on the problem domain. The current Southwest Center for Environmental Research and Policy project provides an obvious starting point, which is sufficiently broad to be challenging but provides some limits to make it feasible. It suggests what types of questions or problems the modeling environment should be able to address. It also suggests instances of such problems and collections of data for common locations on which various models could be used. The difference from the current agenda is that instead of focusing different teams on different problems in order to come up with complementary models, we should establish teams that are working on substitutable but distinctly different models that can be applied to the same datasets and the same problems. The modeling environment should be able not only to link models as complements but to contrast models as replicates.

In the short run, one modeling environment will be sufficient. In the longer run, however, we should also encourage researchers to compete in creating modeling environments for specified problems and questions. Given requirements for a set of modeling capabilities, a set of questions and problems, and a set of locations with data, research teams developing computer modeling environments can measure their success against these requirements. A team can devise a modeling environment and test whether it meets the requirements by attracting other teams to use it in running their models. Given common study sites, modelers can expect to run their models on more than one data set and designers of modeling environments can expect to attract more than one model of a given phenomena at a given study site. A modeling team can devise a model and show its contribution in combination or competition with other models of the same phenomena. Field scientists can devise questions or problems that challenge modeling environments or models.

A research agenda framed and initiated in this way is effective because it creates a game, a conversation, a framework in which research teams can identify opportunities, assess the significance of their contributions, and link their own work to a cumulative enterprise with other researchers. There is motivation both to discover new knowledge about how systems work and to devise new technologies for improving decisions in real situations.

References

Anas, Alex, Richard J. Arnott, and Kenneth A. Small. 1998. Urban spatial structure. *Journal of Economic Literature* 36 (3):1426–1464.

Brill, E. Downey Jr., Shoou-Yuh Chang, and Lewis D. Hopkins. 1982. Modeling to generate alternatives: The HSJ approach and an illustration using a problem in land use planning. *Management Science* 28 (3):221–235.

Brill, E. Downey Jr., John M. Flach, Lewis D. Hopkins, and S. Ranjithan. 1990. MGA: A decision support system for complex, incompletely defined problems. *IEEE Transactions on Systems, Man, and Cybernetics* 20 (4):745–757.

Foot, David. 1981. *Operational urban models*. London: Methuen.

Hopkins, Lewis D. 1979. Quadratic versus linear models for land use plan design. *Environment and Planning A* 11 (3):291–298.

———. 1977a. Land use plan design: Quadratic assignment and central facility models. *Environment and Planning A* 9 (6):625–642.

———. 1977b. Methods for generating land suitability maps: A comparative evaluation. *Journal of the American Institute of Planners* 43 (4):386–400.

Hopkins, Lewis D., E. Downey Brill Jr., and Benedict Wong. 1982. Generating alternative solutions for dynamic programming models of water resources problems. *Water Resources Research* 18 (4):782–790.

Hopkins, Lewis D., and Marc Los. 1979. Location-allocation algorithms for land use plan design with fixed and substitutable interactions. *Journal of Regional Science* 19 (3):345–361.

Lai, Shih-Kung, and Lewis D. Hopkins. 1995. Can decision makers express multiattribute preferences using AHP and MUT? An Experiment. *Environment and Planning B: Planning and Design* 22 (1):21–34.

———. 1989. The meanings of tradeoffs in multi-attribute evaluation methods: A comparison. *Environment and Planning B: Planning and Design* 16 (2):155–170.

Lee, Insung, and Lewis D. Hopkins. 1995. Procedural expertise for efficient multiattribute evaluation: A procedural support strategy for CEA. *Journal of Planning Education and Research* 14 (4):255–268.

Verma, Niraj. 1998. *Similarities, connections, and systems: The search for a new rationality for planning and management*. Lanham, Md.: Lexington Books.

Waddell, Paul. 2000. A behavioral simulation model for metropolitan policy analysis and planning: Residential location and housing market components of UrbanSim. *Environment and Planning B: Planning and Design* 27 (2):247–263.

Wegener, Michael. 1994. Operational urban models: State of the art. *Journal of the American Planning Association* 60 (1):17–29.

Westervelt, James. 2001. *Simulation modeling for watershed management*. New York: Springer-Verlag.

Section II: Ecologic Processes and Their Land Use Implications

How We Will Grow: Baseline Projections of California's Urban Footprint Through the Year 2100

John D. Landis and Michael Reilly

1.0 Introduction

By 2020, most forecasters agree, California will be home to between 43 and 46 million residents—up from 35 million today. Beyond 2020, the size of California's population is less certain. Depending on the composition of the population, and future fertility and migration rates, California's 2050 population could be as little as 50 million or as much as 70 million. One hundred years from now, if present trends continue, California could conceivably have as many as 90 million residents.

Where these future residents will live and work is unclear. For most of the 20th century, two-thirds of Californians have lived south of the Tehachapi Mountains and west of the San Jacinto Mountains—in that part of the state commonly referred to as Southern California. Yet most of coastal Southern California is already highly urbanized, and there is relatively little vacant land available for new development. More recently, slow-growth policies in Northern California and declining developable land supplies in Southern California are squeezing ever more of the state's population growth into the San Joaquin Valley.

How future Californians will occupy the landscape is also unclear. Over the last fifty years, the state's population has grown increasingly urban. Today, nearly 95 percent of Californians live in metropolitan areas, mostly at densities less than ten persons per acre. Recent growth patterns have strongly favored locations near freeways, most of which where built in the 1950s and 1960s. With few new freeways on the planning horizon, how will California's future growth organize itself in space? By national standards, California's large urban areas are already reasonably dense, and economic theory suggests that densities should increase further as California's urban regions continue to grow. In practice, densities have been rising in some urban counties but falling in others.

These are important issues as California plans its long-term future. Will California have enough land of the appropriate types and in the right locations to accommodate its projected population growth? Will future population growth consume ever greater amounts of irreplaceable resource lands and habitat? Will jobs continue decentralizing, pushing out the boundaries of metropolitan areas? Will development densities be sufficient to support mass transit, or will future Californians be stuck in perpetual gridlock? Will urban, resort, and recreational growth in the Sierra Nevada and Trinity Mountain regions lead to the over-fragmentation of precious natural habitat? How much water will be needed by California's future industries, farms, and residents, and where will that water be stored? Where should future highway, transit, and high-speed rail facilities and rights-of-way be located? Most of all, how much will all this growth cost, both economically and in terms of changes in California's quality of life?

Clearly, the more precise our current understanding of how and where California is likely to grow, the sooner and more inexpensively appropriate lands can be acquired for purposes of conservation, recreation, and future facility siting. Similarly, the more clearly future urbanization patterns can be anticipated, the greater our collective ability to undertake sound city, metropolitan, rural, and bioregional planning.

Consider two scenarios for the year 2100. In the first, California's population would grow to 80 million persons and occupy the landscape at an average density of eight persons per acre, the current statewide urban average. Under this scenario and assuming that 10 percent of California's future population growth would occur through infill—that is, on existing urban land—California's expanding urban population would consume an additional 5.06 million acres of currently undeveloped land. As an alternative, assume the share of infill development were increased to 30 percent and that new population were accommodated at a density of about 12 persons per acre—which is the current average density of the city of Los Angeles. Under this second scenario, California's urban population would consume an additional 2.6 million acres of currently undeveloped land. While both scenarios accommodate the same amount of population growth and generate large increments of additional urban development—indeed, some might say even the second scenario allows far too much growth and development—the second scenario is far kinder to California's unique natural landscape.

This paper presents the results of a series of baseline population and urban growth projections for California's 38 urban counties through the year 2100. Presented in both map and table form, these projections are based on extrapolations of current population trends and recent urban development trends. The next section, **Approach**, outlines the methodology and data used to develop the various projections. Section three, **Forecasts**, reports on the forecasts derived from our analyses. Section four, **Baseline Scenario**, reviews the projections themselves. Section five, **Baseline Impacts**, quantitatively assesses the impacts of the baseline projections on wetland, farmland, and habitat loss. The final section underlines key assumptions and caveats and offers some concluding comments.

2.0 Approach

Developing short-term forecasts in a state as diverse and fluid as California is a difficult proposition. Developing long-term forecasts, whether for 20, 50, or 100 years, are harder still. Developing long-term forecasts that are spatially explicit—that project which lands are likely to be developed and which are not—is closer to art than science.

At a conceptual level, our forecasting methodology is actually quite simple (Figure 1). We begin by calibrating a spatial-statistical model of historical development patterns spanning the years 1988 to 1998 (Step A). The calibrated model parameters are then used with contemporary spatial data to generate a development probability surface describing the likelihood that particular undeveloped sites will subsequently be developed (Step B). This is the *where* part of the equation. *When* development happens is a function of state and county population growth pressures (Step 1), the share of population accommodated through infill development (Step 2), and the density at which development occurs (Step 3). Projected population growth, net of infill, is then allocated to allowable development sites in order of their projected development probability (from Step B) at a designated development density. Once a future allocation has been completed (e.g., for the 2000–2020 period), infill rates, densities, and development probabilities are updated to reflect any intervening changes. The model is then run again (Steps 1 through 5) for subsequent periods. Alternative futures or scenarios can be tested any number of ways. Different growth increments can be postulated. Allocation densities can be adjusted up or down for regions, counties, or individual jurisdictions. More or less land can be excluded from development *a priori* through the specification of additional or fewer exclusion conditions. By changing a few input settings for example, it is possible to compare a "business-as-usual" scenario involving the extension of recent development trends to an "environmental protection" scenario in which development densities are increased and development is precluded from occurring on farmlands, wetlands, hillsides and sensitive habitats or to an "infrastructure investment" alternative in which new highways and transit systems are constructed.

2.1 Growth Model Calibration

Before a statistical model can be used to generate future projections, it must be calibrated. With nonspatial models, this usually involves fitting a line or curve to historical data. With spatial data, this involves developing equations and estimating parameters that are sensitive to locational as well as nonlocational influences. In this case, the model being calibrated relates changes in the development status of particular sites between 1988 and 1998—measured as a matrix of one-hectare

Figure 1.

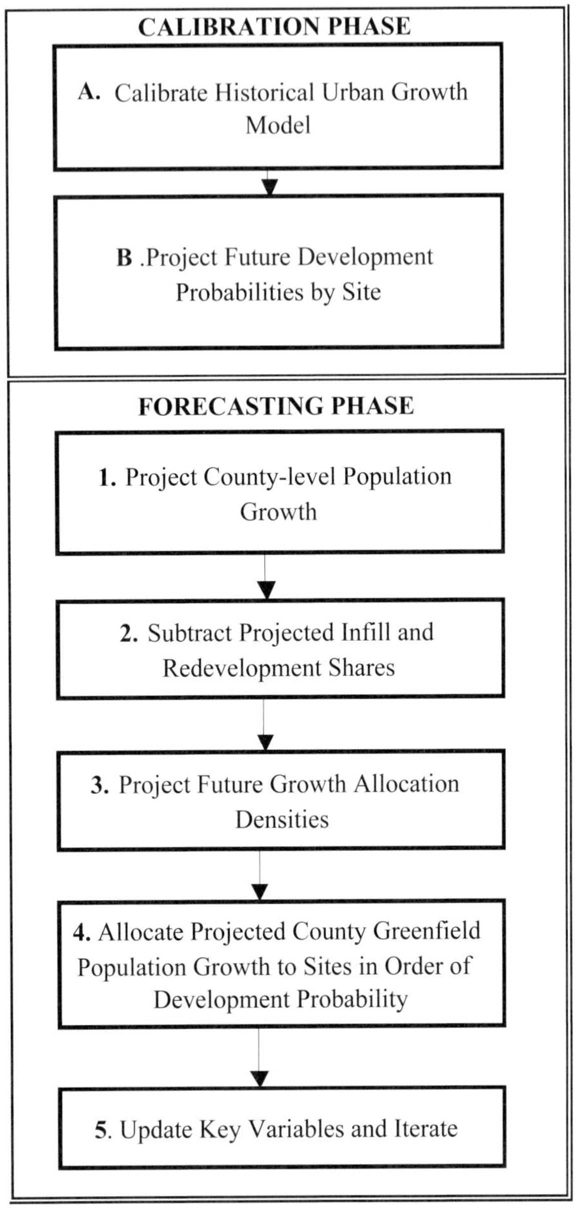
CALIBRATION PHASE
A. Calibrate Historical Urban Growth Model
B .Project Future Development Probabilities by Site
FORECASTING PHASE
1. Project County-level Population Growth
2. Subtract Projected Infill and Redevelopment Shares
3. Project Future Growth Allocation Densities
4. Allocate Projected County Greenfield Population Growth to Sites in Order of Development Probability
5. Update Key Variables and Iterate

grid cells—and their various physical, locational, and administrative characteristics. As with all statistical models, the estimated parameters describe the relationship between a set of independent or explanatory variables and a single dependent variable:

$$\text{Prob [Undeveloped site}_I > \text{Developed]} = f(X1_I, X2_I, \ldots\ldots Xn_I)$$

The dependent variable in this case is the change in development status between 1988 and 1998 of all potentially developable *sites*, measured as a matrix of one-hectare grid cells. Sites that were undeveloped in 1988 and remained that way through 1998 are assigned a value of "0." Sites, which were developed between 1988 and 1998, are assigned a value of "1." Land-use change information was obtained from the California Farmland Mapping and Monitoring Program (CFMMP), a division of the California Department of Conservation. Through a combination of remote-sensing and local ground-truthing, the CFFMP conducts detailed biannual land cover inventories for 38 California counties. CFMMP data is generally accurate down to the one-hectare level.

The *X*s, or independent variables, are those attributes thought most likely to affect each site's conversion from nonurban to urban use. Independent variables can include physical site characteristics, locational and economic characteristics, the characteristics of nearby sites, and policy and administrative characteristics such as the presence of a local growth control measures. Once measured, the dependent and independent variables are matched spatially using GIS.

Because the dependent variable is categorical rather than continuous, the model is estimated using logistical regression, also known as *logit*, rather than linear regression. Model parameters are estimated using a maximum-likelihood procedure in which the error terms are presumed to follow a Weibull distribution. In this case, because the dependent variable takes on just two categorical values (e.g., indicating either a change in land use or no change in land use), the type of logit model presented above is known as a binomial logit model.[1]

The use of small grid-cells as surrogates for development sites exacerbates a problem known as *spatial autocorrelation*. Spatial autocorrelation refers to the fact that adjacent or nearby objects tend to influence each other. Some types of spatial autocorrelation are legitimate, as in the case of the rancher who observes his next door neighbor selling to a developer and is influenced to do the same. Other types of spatial autocorrelation are artifacts, generated by the choice of the spatial unit of analysis. If, as in the current case, one-hectare grid cells are used to record land-use change events, then any land-use changes larger than one hectare will be recorded as multiple, adjacent events. The resulting over-counting of land-use change will tend to bias the results of any statistical models calibrated on the basis of those changes. There is as yet no commonly available modeling package that corrects for spatial autocorrelation. As noted below, we attempt to do so through the explicit inclusion of *neighborhood*-level independent variables.

Four types of measures were included as independent variables:

1. *Demand Variables,* measure the demand for sites as a function of their accessibility to job opportunities and job growth, as well as local income

levels. Two demand variables are included in each model: JOB_ACCESS90, which measures the number of jobs within 90 minutes of a given grid-cell and assumes travel times of 50 mph on freeways and limited access roads and 25 mph on local roads; and INC_RATIO90, which is the ratio of community median household income to county median household income. All else being equal, we would expect sites with superior job accessibility more likely to be developed and sites in upper-income communities less likely to be developed.

2. *Own-site Variables*, measure the physical and land-use characteristics of each grid-cell as determinants of its development potential. Four own-site variables are modeled: FRWY_DISTSQ, a measure of the squared distance from each site to the nearest freeway; PRIME_FARM, a dummy variable that indicates whether the site is classified as prime farmland by the CFMMP; SLOPE, the average percentage slope of each site; and FLOOD, a dummy variable indicating whether the site falls within the Federal Emergency Management Agency (FEMA)-designated 100-year flood zone. Based on cost and market considerations, we would expect sites near freeways more likely to be developed and sites classified as prime farmland or in flood zones less likely to be developed. Similarly, based on the higher cost of building on steep slopes, we would expect the probability of a site being developed to be inversely proportional to its slope.[2]
3. *Adjacency and Neighborhood Variables*, summarize the environmental and land-use characteristics of adjacent and neighboring grid-cells. Four neighborhood variables are modeled: SLOPE_1KM, the average slope of the cells within one kilometer of each subject site; SLOPE_2-3KM, the average slope of sites within the two-to-three kilometer ring around each subject site; FLOOD_1KM, the share of sites within one kilometer of the subject site that are located in the FEMA 100-year flood zone; and FLOOD_2-3KM, the share of sites within the two-to-three kilometer ring around each subject site. Including these variables in the model offers two benefits. It allows the characteristics of adjacent and neighboring sites to affect the development of subject sites (e.g., a flat site surrounded by steep slopes is presumed to be less likely to be developed), as well as reducing parameter bias due to potential spatial autocorrelation.
4. *Regulatory and Administrative Variables*, are intended to capture the development encouraging or constraining effects of different land-use policies and regulations. With respect to land use policy, the dummy variable, IN_CITY, denotes whether or not a site is located within an incorporated city. Most California jurisdictions provide more services and a higher level of services within incorporated cities. Many California cities and counties work collaboratively to encourage city-centered development and discourage growth in unincorporated areas. We would thus expect sites located within incorporated cities more likely to be developed than unincorporated county lands. A second set of dummy variables, one

for each county, is included to reflect inter-county differences in land-use regulation.

The calibration sample consists of all one-hectare sites in a county that were undeveloped as of 1988, that were not publicly-owned (and therefore could be developed), that had a slope of less than fifteen percent, and that were within 15 kilometers (nine miles) of a major highway or existing urban development .

To better account for systematic regional variations, we tested separate models for Southern California, Northern California, the Sacramento region, and the San Joaquin Valley. The Northern California study area includes the nine counties of the Bay Area (Alameda, Contra Costa, Marin, Napa, San Francisco, San Mateo, Santa Clara, Solano, and Sonoma), as well as five neighboring counties (Monterey, San Benito, San Joaquin, Santa Cruz, and Stanislaus) that now fall within commuting range of the Bay Area. The Southern California study area includes Imperial, Los Angeles, Orange, Riverside, San Bernardino, San Diego, San Luis Obispo, Santa Barbara, and Ventura counties. The Sacramento region includes Sacramento County as well as Yolo, Sutter, Yuba, El Dorado, and Placer counties. The San Joaquin Valley includes Kern, Fresno, Madera, Merced, Kings, and Tulare counties.

The results of the four regional models are presented in Table 1. Overall, the four models fit the data extremely well, explaining more than 95 percent of urban land use between 1988 and 1998 in their respective regions.

We report both the standardized parameter estimate and the odds-ratio for each independent variable. Except where noted, all of the parameter estimates are statistically significant, and most are of the expected signs. The importance of particular factors varies by region.

Among Southern California counties, the factors that most increased the likelihood of site development most during the 1990s were freeway proximity (FRWY_DISTSQ), job accessibility (JOB_ACCESS), being located in a city (IN_CITY), and being located in Santa Barbara or San Diego counties. Steeply-sloped sites were less likely to be developed than flatter sites, and prime farmlands were somewhat less likely to be developed. Reflecting NIMBY pressures. Sites in upper-income communities were significantly less likely to be developed than sites in middle- or lower-income communities. All else being equal, sites in San Bernardino County were less likely to be developed than sites elsewhere in Southern California.

Among Northern California counties, the factors that most increased the likelihood of site development during the 1990s were freeway proximity (FRWY_DISTSQ), being located in a city (IN_CITY), and being located in Napa, Sonoma, Santa Cruz, Monterey, and Stanislaus counties. Compared to Southern California, steeply sloped sites and prime farmlands in Northern California were far less likely to be developed than flatter and less fertile locations. Sites in Solano County were less likely to be developed than sites elsewhere in the Bay Areas, as were sites in and around flood zones. Accessibility to jobs, while a positive influence on development, was far less significant in northern California than southern California. Surprisingly, sites in wealthy communities in northern California were

Table 1. Logistic Regression Model of 1988–98 Site-level Land Use Changes in Southern California, the Bay Area, the Sacramento Region, and the Southern San Joaquin Valley*

Dependent Variable : Probability of site-level land use change, 1988–1998		**Southern California region**		**Northern California region**		**Sacramento region**		**S. San Joaquin Valley**	
Independent Variables	Abbreviation	Standardized Coefficient	Prob. Level	Standardized Coefficient	Prob. Level	Standardized Coefficient	Prob. Level	Standardized Coefficient	Prob. Level
Dummy Variable [Within incorporated city]	IN_CITY	0.185	0.00	0.179	0.00	0.017	0.00	0.146	0.00
Distance to freeway(km) - squared	FRWY_DISTSQ	-0.297	0.00	-0.305	0.00	-0.231	0.00	-0.0018	0.80
Regional job accessibility as of 1990	JOB_ACCESS	0.180	0.00	0.073	0.00	0.102	0.00	0.2070	0.00
Ratio of 1990 City-to-Region Median HH Income	INC_RATIO90	-0.032	0.00	0.005	0.00	-0.007	0.01	not entered	
Dummy Variable [CFMMP-designated Prime farmland]	PRIME	-0.007	0.02	-0.045	0.00	-0.131	0.00	-0.0018	0.68
Dummy Variable [FEMA Flood zone]	FLOOD	not entered		-0.023	0.01	-0.151	0.00	-0.038	0.00
Flood zone 1x Nbr. Percent	FLOOD_1X	not entered		-0.060	0.00	0.102	0.00	-0.024	0.02
Flood zone 2-3x Nbr. Percent	FLOOD_2X	not entered		-0.097	0.00	not entered		not entered	
Site slope	SLOPE	-0.031	0.00	-0.033	0.00	0.038	0.00	0.0127	0.38
Avg. slope of 1x adjacent sites	ADJ_SLOPE	-0.192	0.00	-0.291	0.00	-0.187	0.00	-0.455	0.00
Avg. slope of 1x-2x adjacent sites	NEIGH_SLOPE	0.031	0.00	-0.111	0.00	0.168	0.00	0.209	0.00
Intercept		-4.695	0.00	-5.349	0.00	-5.655	0.00	-6.463	0.00

* Individual county dummy variables are not reported

actually more likely to be developed than sites in poorer communities—the opposite situation than in southern California.

Among Sacramento area counties, the factors that most affected the likelihood of site development during the 1990s were freeway proximity (FRWY_DISTSQ) and whether or not the site was located in a flood zone (FLOOD), or on prime farmland (PRIME). Sites near freeways were much more likely to be developed, while flood zone and prime farmland sites were much less likely to be developed. Job accessibility (JOB_ACCESS) was also an important influence. Sites located in incorporated cities were only marginally more likely to be developed than unincorporated sites—a finding in contrast to the southern and northern California regions, where development strongly favored incorporated sites. The effect of community income on development activity, while negative, was also slight. Compared to sites in Sacramento County, sites in Nevada County were much more likely to be developed between 1988 and 1998, while sites in Yolo County were much less likely to be developed. Sites in El Dorado and Placer counties were marginally more likely to be developed, and sites in Yuba County were marginally less likely to be developed.

Among counties in the San Joaquin Valley—including Fresno, Kern, Kings, Madera, Merced, and Tulare—the two factors that most affected the likelihood of site development during the 1990s were regional job accessibility (JOB_ACCESS) and whether the site was located in an incorporated city. Sites with good accessibility to jobs were much more likely to have developed, as were sites in incorporated cities. As in the Sacramento region, hillside sites were slightly more likely to be developed than valley sites. All else being equal, freeway accessibility had a much smaller effect on site developability in the southern San Joaquin Valley than elsewhere in the state. On the negative side, flood zone sites and sites located on prime farmland were less likely to be developed than other less environmentally sensitive sites, although the differences were not large. Compared to comparable sites in Fresno County, sites in Kings and Tulare counties were somewhat more likely to be developed between 1988 and 1998, while sites in Madera and Kern counties were less likely to have been developed. Sites in Merced County were marginally more likely to have been developed than sites in Fresno County.

Once estimated, the various model parameters can be used to generate development probability scores for all remaining undeveloped sites. Use of these scores for forecasting requires assuming that the particular factors that influenced development in the recent past will continue to do so in the future and in the same combination. To the extent that the future brings no large technological or land-use policy changes, or significant shifts in household and business location preferences, the assumption that future land development trends will follow those of the past may be quite reasonable. On the other hand, to the extent that land-use preferences, policies, and technologies *all change*, the usefulness of models calibrated using historical data is obviously reduced.

2.2 Patterns of Job Growth

Depending on the region, job accessibility is either the second, third, or fourth most important determinant of urban growth patterns in California (see Table 1). Having long-term, accurate, and spatially disaggregate job projections is thus a prerequisite to developing accurate growth scenarios. This is easier said than done. The phrase "long-term, accurate, and spatially disaggregate job projections" is an oxymoron. Economies are by their very nature interdependent and unpredictable. Most available employment projections are therefore short-term and subject to constant revision. In terms of space, most job projections are undertaken at the MSA or county level—as much for reasons of data availability as modeling capability.

Our approach to forecasting jobs and job accessibility is a little different. Rather than generating separate sectoral and county-level job forecasts and then aggregating them into a single regional total—as is common practice—we started with the presumption that there is a more or less regular relationship between the size of a region's population and its employment base.[3] Acceptance of this assumption means that one can use believable regional population projections as a starting point for developing serviceable regional employment projections.

The major challenge for our purposes is not to project the total number of new jobs. Rather, it is to figure out where in each region those new jobs are most likely to locate. Fortunately, the long-term spatial trend is quite clear. Broadly speaking, we expect jobs in California to continue their historical pattern of intra-metropolitan decentralization. Prior to 1950, most basic[4] jobs in the U.S. economy were located in urban cores. Since 1950, job growth has increasingly favored suburban communities over urban cores. Since 1980, almost all net basic job growth has occurred outside traditional central cities[5]. The Los Angeles region and more recently the San Francisco Bay Area have been national leaders in the trend toward increased job decentralization.

To project future job decentralization, we start by comparing the 1990 and 2000 spatial distribution of jobs in each California metropolitan region. Employment estimates were obtained from multiple sources, most notably the Southern California Association of Governments (SCAG), the Association of Bay Area Governments (ABAG), the Sacramento Area Council of Governments (SACOG), and SANDAG, the San Diego Association of Governments. Job estimates for the South San Joaquin Region were obtained from each county council of governments. Job totals for 1990 and 2000 were then mapped by city and CDP. Next, 10-kilometer-wide rings were generated outward from each regional center and used to count the number of job centers and total number of jobs in each ring. Next, a spatial shift-share model was applied to decompose 1990–2000 city and CDP job changes into three components:

1. *A Regional Growth Component (RGC)* calculated as the percent change in regional jobs between 1990 and 2000:

   (2000 Regional Jobs/1990 Regional Jobs)

The larger the regional growth component, the more vital the entire regional economy.

2. *A Ring Change Component (RCC)* calculated as the difference between the 1990–2000 percent change in jobs in each ring and the Regional Growth Component:

   (2000 Ring $Jobs_i$/1990 Ring $Jobs_i$) - (2000 Regional Jobs/1990 Regional Jobs)

   Rings with RCC values greater than zero added jobs at a faster rate than the region as a whole. Rings with RCC values less than zero added jobs at a slower rate than the region as a whole.

3. *A Local Change Component (LCC)* calculated as the difference between the 1990–2000 percent change in jobs in each city or CDP and the 1990–2000 percent change in jobs in its respective ring:

   (2000 Local $Jobs_j$/1990 Local $Jobs_j$) - (2000 Ring $Jobs_i$/1990 Ring $Jobs_i$)

   Localities with LCC values greater than zero added jobs at a faster rate than their rings. Localities with LCC values less than zero added jobs at a slower rate than their rings.

The results of this analysis show that the outer rings in each metropolitan area added jobs at a faster rate during the 1990s than the inner rings. The inner-most Southern California ring actually lost jobs between 1990 and 2000, the only such ring to do so. Among the four regions profiled, the rate of inner-ring job growth was highest in northern California, while the rate of outer-ring job growth was greatest in southern California.

2.3 Forecasting Procedures

As previously noted, forecasting and scenario-building involves five distinct steps:

1. *Project county-level population growth through 2100.* County population projections for the year 2020 and 2040 were obtained from the California Department of Finance, Population Research Unit.[6] These projections were used to estimate annualized population growth rates (by county) spanning the periods 2000–2040 and 2020–2040. Projected forward, these growth rates were used in turn to forecast county population totals for the years 2050 and 2100.
2. *Subtract projected infill and redevelopment shares.* A significant share of projected population growth will occur within the existing urban footprint in the form of infill or redevelopment. Infill shares tend to rise over time as remaining greenfield areas are used up and as developers reconsider previously passed-over infill lands. A cross-sectional regression model was developed relating current county infill shares to remaining greenfield land supplies. This model was then used to project future infill and greenfield population shares for the years 2020, 2050, and 2100.

3. *Project future allocation densities.* The amount of greenfield land consumed by future population growth will depend both on the magnitude of growth and gross density. Marginal gross densities—that is the gross densities of new development—were estimated for each county by dividing the change in the population between 1988 and 1998 by the change in urbanized land area for the same period. Theory suggests that densities should rise as available greenfield lands are used up, as developers seek to use remaining lands more intensely. A cross-sectional regression model was developed relating 1988–98 marginal densities to remaining greenfield land supplies. This model was then used to project future allocation densities by county for the years 2020, 2050, and 2100.
4. *Allocate projected greenfield population growth to undeveloped sites in each region in order of development probability.* Starting with the hectare-scale development probability scores derived above, a series of exclusion conditions were developed identifying which sites were to be precluded from development. Projected population growth (from Step 2) for the period 2000–2020 is then allocated to sites at projected densities (from Step 3) in order of development probability (from high to low) subject to any exclusion conditions.
5. *Update key variables to reflect projected employment growth and allocated population growth.*

Steps 4 and 5 are iterated for the periods 2020–2050, and 2050–2100. Thanks to the analytical power of GIS, different forecasting steps can be undertaken at different spatial scales and then reconciled. Population growth, greenfield shares, and allocation densities, for example, are all identified and projected (Steps 1,2, and 3) at the county level. Development probability scores, on the other hand, are estimated for individual one-hectare sites, accounting for differences among counties and regions. Employment projections, an input into the allocation procedure (Step 4) are developed for individual job centers. Distance to city boundaries, another input into the allocation procedure, is estimated and updated for incorporated cities.

The following section discusses the results of each of the above procedures in greater detail.

3.0 Forecasts

3.1 Population Projections—Huge Growth Ahead

Forecasters project large area population growth in one of two ways: by extrapolating a single long-term population growth trend or by decomposing that trend into its two component parts—natural increase and net migration—and then projecting those. The California Department of Finance (DoF), which is required by state law to develop forty-year county-level population projections, takes the latter approach.

Natural increase is the difference between births and deaths and generally follows fertility rate trends. Following accepted demographic practice, DoF identifies natural increase and fertility rate trends by age cohort and race and ethnicity. Fertility rates vary as well by immigration status and length of residency, although not always in predictable ways. Net migration measures the difference between in- and out-migration and for the most part follows job growth trends—rising when the economy is booming and falling when it is in recession. Like fertility rates, net-migration rates vary by population age—higher for young adults, migration rates typically decline with age. County and, to a lesser extent, state migration rates also vary with the relative cost of living, as new migrants are often shunted into counties with more affordable housing. Some of these complications wash-out at the state level, but serve to make county-level forecasting all the more complicated.

The Department of Finance's E-6 county-level population projections were developed using the cohort component method described above. These projections were then used to calculate composite annual growth rates by county for the years 2000–2040. These rates, vary from a high of 3.0 percent per year for Imperial County to a low of -.4 percent per year for San Francisco County. Annualized 2020–2040 growth rates are somewhat lower, and range from a high of 2.65 percent, also for Imperial County, to a low of -.5% percent for San Francisco County.

Given that high growth rates are rarely sustainable over the long term and that the growth rates of low-growth counties located in high-growth states tend to pick-up over time we averaged each county's 2000–2040 and 2020–2040 growth rates with those of the state as a whole.

We project each county's population forward to 2050 and 2100, based first on the lower 2020-2040 combined rate and then second on the higher 2000–2040 combined rate. Table 2 provides estimates and forecasts of California population by county, subregion, and region. Based on this method, California's largest county, Los Angeles, will grow from 10 million people in 2000 to 15.5 million by 2050. The populations of Riverside, San Bernardino, and San Diego counties will each exceed 5 million by 2050. The population of Orange County will grow from 2.8 million in 2000 to more than 4.5 million in 2050. Elsewhere, the 2050 population of the largest county in northern California, Santa Clara, will be just under 3 million. With a 2050 population of 2.4 million, Sacramento County will be the most populous in the Central Valley. Added up, the total 2050 population of California's 58 counties will exceed 66 million!

Projecting further forward to the year 2100 presents additional challenges. Given the immense size of California's population, even the lower 2020–2040 growth rates are likely to be unsustainable over time. To better reflect the natural tendency for growth rates to decline as the population increases, we reduced both the lower 2020–2040 composite growth rate and the higher 2000–2040 composite growth rate by 50 percent before applying them to the 2050–2100 period.

Table 2. Population Projections by County, Subregion, and Region for 2000, 2020, 2050, and 2100

Major Region	County	Subregion	Population Estimates and Forecasts				County Share of Regional Change		
			2000*	**2020F***	**2050F**	**2100F**	**2000–2020**	**2020–2050**	**2050–2100**
Southern California	Los Angeles	Central	9,838,861	11,575,693	15,497,560	20,400,280	**28.7%**	**30.5%**	**31.3%**
	Imperial	South	154,549	298,700	612,914	1,000,884	2.4%	2.4%	2.5%
	Orange	South	2,833,190	3,431,869	4,535,936	5,932,517	9.9%	8.6%	8.9%
	San Diego	South	2,943,001	3,917,001	5,831,574	8,097,302	16.1%	14.9%	14.4%
	Subregional Total		**5,930,740**	**7,647,570**	**10,980,424**	**15,030,702**	28.4%	25.9%	25.8%
	Riverside	Inland Empire	1,570,885	2,773,431	5,335,081	8,431,480	19.9%	19.9%	19.7%
	San Bernardino	Inland Empire	1,727,452	2,747,213	4,983,011	7,644,175	16.9%	17.4%	17.0%
	Subregional Total		**3,298,337**	**5,520,644**	**10,318,093**	**16,075,656**	**36.8%**	**37.2%**	**36.7%**
	Santa Barbara	North	412,071	552,846	905,294	1,318,823	2.3%	2.7%	2.6%
	Ventura	North	753,820	981,565	1,456,134	2,018,255	3.8%	3.7%	3.6%
	Subregional Total		**1,165,891**	**1,534,411**	**2,361,429**	**3,337,078**	**6.1%**	**6.4%**	**6.2%**
	REGIONAL TOTAL		**20,233,829**	**26,278,318**	**39,157,506**	**54,843,715**	**100.0%**	**100.0%**	**100.0%**
Northern California	Alameda	Central	1,470,155	1,793,139	2,287,126	2,938,378	17.9%	15.9%	16.1%
	Contra Costa	Central	931,946	1,104,725	1,394,436	1,782,151	9.6%	9.3%	9.6%
	San Francisco	Central	792,049	750,904	710,034	785,565	-2.3%	-1.3%	1.9%
	San Mateo	Central	747,061	855,506	1,044,065	1,312,014	6.0%	6.1%	6.6%
	Santa Clara	Central	1,763,252	2,196,750	2,884,875	3,760,965	24.0%	22.1%	21.6%
	Subregional Total		**5,704,463**	**6,701,024**	**8,320,538**	**10,579,072**	**55.3%**	**52.0%**	**55.7%**
	Marin	North	248,397	268,630	325,152	406,920	1.1%	1.8%	2.0%
	Napa	North	127,084	157,878	214,934	285,317	1.7%	1.8%	1.7%
	Solano	North	399,841	552,105	789,742	1,074,736	8.4%	7.6%	7.0%
	Sonoma	North	459,258	614,173	845,837	1,129,343	8.6%	7.4%	7.0%
	Subregional Total		**1,234,580**	**1,592,786**	**2,175,666**	**2,896,317**	**19.9%**	**18.7%**	**17.8%**
	Monterey	South	401,886	575,102	1,006,978	1,517,431	9.6%	13.9%	12.6%
	San Benito	South	51,853	82,276	133,208	192,948	1.7%	1.6%	1.5%
	San Luis Obispo	South	254,818	392,329	617,709	882,227	7.6%	7.2%	6.5%
	Santa Cruz	South	260,248	367,196	572,017	812,597	5.9%	6.6%	5.9%
	Subregional Total		**968,805**	**1,416,903**	**2,329,912**	**3,405,203**	**24.9%**	**29.3%**	**26.5%**
	REGIONAL TOTAL		**7,907,848**	**9,710,713**	**12,826,116**	**16,880,592**	**100.0%**	**100.0%**	**100.0%**

***Source: Department of Finance, State of California**

Major Region	County	Subregion	Population Estimates and Forecasts				County Share of Regional Change		
			2000*	2020F*	2050F	2100F	2000–2020	2020–2050	2050–2100
San Joaquin Valley	Merced	North	215,256	319,785	537,166	792,667	6.1%	6.3%	6.3%
	San Joaquin	North	579,172	884,375	1,454,089	2,122,660	17.9%	16.5%	16.5%
	Stanislaus	North	459,025	708,950	1,160,376	1,690,026	14.6%	13.1%	13.0%
	Subregional Total		**1,253,453**	**1,913,110**	**3,151,631**	**4,605,353**	**38.6%**	**35.9%**	**35.8%**
	Fresno	South	811,179	1,114,403	1,753,356	2,503,297	17.8%	18.5%	18.5%
	Kern	South	677,372	1,073,748	1,919,849	2,923,829	23.2%	24.5%	24.7%
	Kings	South	126,672	186,611	309,815	454,484	3.5%	3.6%	3.6%
	Madera	South	126,394	224,567	411,713	635,019	5.8%	5.4%	5.5%
	Tulare	South	379,944	569,896	982,425	1,468,811	11.1%	12.0%	12.0%
	Subregional Total		**2,121,561**	**3,169,225**	**5,377,159**	**7,985,439**	**61.4%**	**64.1%**	**64.2%**
	REGIONAL TOTAL		**3,375,014**	**5,082,335**	**8,528,790**	**12,590,792**	**100.0%**	**100.0%**	**100.0%**
Sacramento	Sacr mento	Central	**1,212,527**	**1,651,765**	**2,409,784**	**3,312,096**	**52.6%**	**56.0%**	**56.1%**
	El Dorado	Foothills	163,197	256,119	381,668	530,209	11.1%	9.3%	9.2%
	Nevada	Foothills	97,020	136,405	185,998	247,103	4.7%	3.7%	3.8%
	Placer	Foothills	243,646	391,245	598,462	842,385	17.7%	15.3%	15.2%
	Subregional Total		**503,863**	**783,769**	**1,166,127**	**1,619,697**	**33.5%**	**28.2%**	**28.2%**
	Sutter	North	82,040	116,408	173,672	241,405	4.1%	4.2%	4.2%
	Yuba	North	63,983	84,610	124,998	172,890	2.5%	3.0%	3.0%
	Subregional Total		**146,023**	**201,018**	**298,670**	**414,295**	**6.6%**	**7.2%**	**7.2%**
	Yolo	West	**164,010**	**225,321**	**341,228**	**477,893**	**7.3%**	**8.6%**	**8.5%**
	REGIONAL TOTAL		**2,026,423**	**2,861,873**	**4,215,809**	**5,823,981**	**100.0%**	**100.0%**	**100.0%**
	Non-Metropolitan Counties		**1,109,741**	**1,515,388**	**2,177,969**	**2,990,087**	**100.0%**	**100.0%**	**100.0%**
	CALIFORNIA		**34,653,395**	**45,448,627**	**66,763,758**	**92,081,030**	**100.0%**	**100.0%**	**100.0%**

***Source: Department of Finance, State of California**

Despite the imposed slowdown in growth rates, California's largest counties will continue to grow. California's largest county, Los Angeles, will grow from 15.5 million people in 2050 to 20.4 million by 2100. The populations of Riverside, San Bernardino, and San Diego counties will each approach or exceed 5 million by 2050 and 7.5 million by 2100. The population of Orange County will grow from 4.5 million in 2000 to nearly 5 million by 2050 and nearly 6 million by 2100. Elsewhere, the 2100 population of the largest in northern California, Santa Clara, will be about 3.8 million. With a 2100 population of 3.3 million, Sacramento County will still be the most populous in the Central Valley. Added up, the total 2100 population of California's 58 counties could very well exceed 92 million!

The huge size of these projections—particularly among southern California counties—clearly indicates the dangers implicit in the long-term use of average annual growth rates. Even so, as large as these projections may seem, they are not unbelievable. California's population in the 1900 was just around 1.5 million. One hundred years later, the state's population stood at nearly 35 million.

3.2 Infill Shares and Growth Densities—Both Will Increase

The location and density of new urban development in California is shaped by two opposing forces. Development has traditionally been attracted to California's coastal areas both for reasons of economics—that's where the ports are—and amenities—the climate along the coast is more moderate. Accordingly, housing and land prices in California have long formed a downward-sloping gradient eastward from the coastal centers of Los Angeles, San Francisco, and San Diego.

As coastal locations become increasingly built-out, developers have moved ever inland. In addition to being less expensive, inland locations have traditionally been less subject to land-use and environmental regulation than their coastal counterparts, making development cheaper and easier.

California was built by developers, and developers are nothing if not opportunistic. Even as they continue their inexorable eastward push, California's developers also continually look back over their shoulders to consider potential infill and redevelopment opportunities. Thus, at the same time that California's coastal metropolitan areas are growing eastward, they are also infilling and redeveloping. And to the extent that infill development tends to occur at higher-than-existing densities, overall urban densities also rise.

At least this is the theory. In practice, local land-use controls and opposition from neighborhood groups often function to make infill and redevelopment proportionately more difficult than greenfield development, thereby breaking the link between growth at the urban fringe, increased infill activity, and rising urban densities. The result is less urban redevelopment and more sprawl.

Figures 2 and 3 graphically present these relationships for 38 predominantly urban counties. Figure 2 compares the share of each county's land area that was urbanized in 1972 with the population density of subsequent new development. As predicted, marginal densities—measured as the change in population divided by change in urban land area—rise with the share of each county's land in urban use. Based on the fitted trend line, for every percent share of each county's land area in urban use in 1972, marginal development densities during the 1972–1996 period rose by .26 persons per acre.

Figure 2.

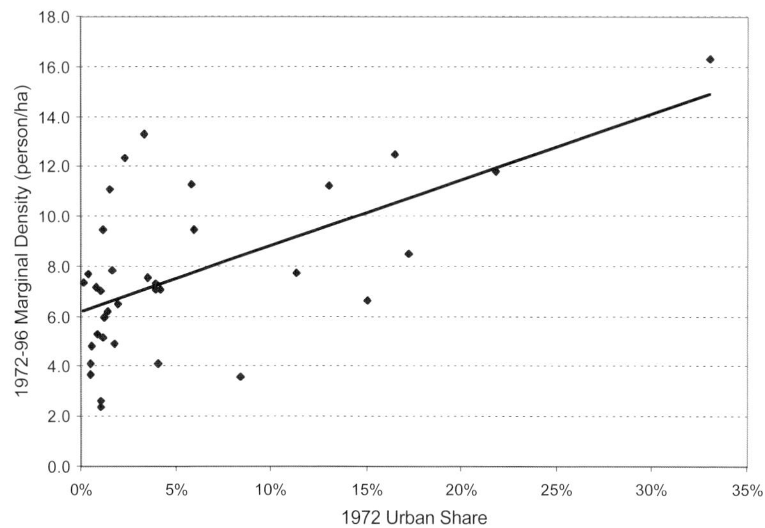

Figure 3.

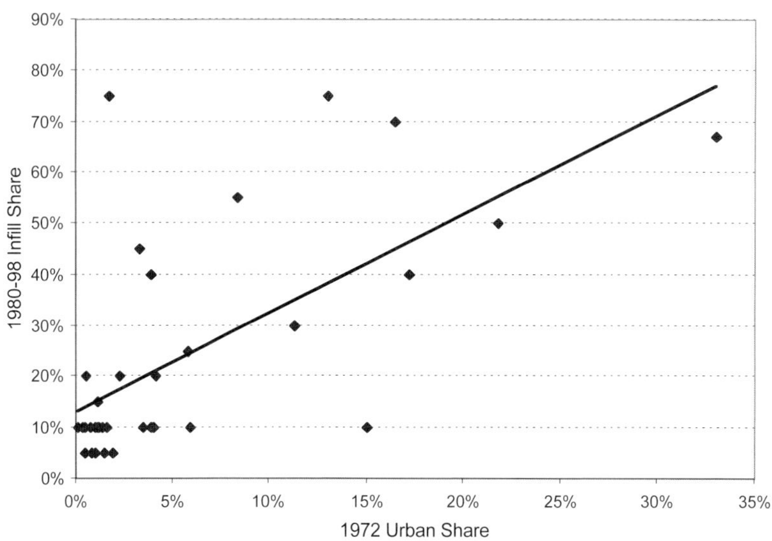

Figure 3 compares the share of each county's land area that was urbanized as of 1972 with the share of new development occurring within the existing urban footprint in the form of infill. As expected, county infill shares rise (and greenfield shares fall) with the share of each county's land in urban use. Based on the fitted trend line, for every percent share of each county's land area in urban use in 1972, the share of subsequent urban development occurring as infill—that is, within the initial 1980 urban footprint—rose by 200 percent.

Neither the density trend line in Figure 2 nor the infill trend line in Figure 3 fits the observed data all that well, a fact confirmed by the middling goodness-of-fit statistics of the estimated regression lines. Some counties, such as Los Angeles, Orange, Santa Clara, San Mateo, and Stanislaus developed at higher densities and with more infill than average. Others, most notably Alameda, Contra Costa, and Sacramento developed at either lower densities or with less infill than expected.

Used with care, these two regression lines can be used to project future development densities and infill shares. In both cases, this involves incorporating additional information:

1. Incremental densities are projected by selecting the maximum of the recent incremental density for each county (denoted by the subscript *i* below) and the regression-based incremental density. This adjustment has the effect of preventing projected incremental densities from falling.

   Projected incremental density$_i$ = MAX [recent incremental density$_i$, regression-based incremental density$_i$]

2. County-level infill growth shares are projected as the average of the current infill share and the maximum of the current infill share and the regression-estimated infill share. This adjustment has the effect of preventing infill shares from either rising too quickly or else falling.

   Projected infill share$_i$ = {Current infill share$_i$ + MAX [current infill share$_i$, regression-based infill share$_i$]} / 2

3. Projected greenfield population growth—that is, the amount of population growth not projected to take the form of infill development—is calculated by multiplying projected population growth for each county by 1.0 minus the projected infill share for that county. The result of this calculation is then multiplied by the projected incremental density to yield an estimate of the amount of additional projected greenfield development.

   Projected Greenfield population growth$_i$ = Population projection$_i$ * [1- projected infill share$_i$]

   Projected greenfield development in acres$_i$ = Projected Greenfield population growth$_i$ * Projected incremental density$_i$] / 2.47

3.3 Updating the Inputs

Projected population growth is allocated to sites during three periods: 1997–2020, 2020 2050, and 2050–2100. Several parameters and data layers are updated prior to each successive allocation round. These include:

1. *Job Accessibility*: A job accessibility measure is calculated for each site based on its proximity via the highway network to all jobs—as located at discrete job centers. As noted in Table 1, this measure is used in the logit model equation used to estimate future site-level development probabilities. Subsequent to each growth allocation, a new set of job accessibility measures is computed for each site based on projected job growth by city or place.
2. *City boundaries*: Because development in California generally favors locations within cities—with some important differences among regions—it is essential to update city boundaries subsequent to each growth allocation and to then estimate development probabilities accordingly. This does not present a problem for newly developed sites within existing city boundaries, but for sites outside existing boundaries, those boundaries must be stretched to accommodate the additional growth. This is done manually. In the most common case, increments of new development adjacent to or nearby existing cities are effectively "incorporated." In rarer cases, small, freestanding increments of new development are treated as unincorporated urban places. In rarer cases still, large, freestanding increments of new development are incorporated as new cities.
3. *Physical features*: The physical features of sites—such as their slope, location in a flood zone, or status as prime farmland—do not change between allocation rounds.
4. *Urban Share*: Subsequent to each allocation round, the share of land area in each county in urban use is updated. The updated urban share is then used to estimate updated incremental development densities and infill shares for the next allocation round.

4.0 The Baseline Scenario

4.1 Building the Baseline Scenario

The function of the Baseline Scenario is to serve as a minimum-change alternative against which future scenarios, which posit more extensive policy, regulatory, or investment interventions, can be compared. More succinctly, the Baseline Scenario assumes continued growth along the lines of past trends and patterns without significant policy change. Among the list of possible policy interventions *not envisioned* in the Baseline Scenario are additional infrastructure projects, additional environmental restrictions on land development, additional conservation and land preservation initiatives, and locally initiated changes in development densities and infill activities.

The Baseline Scenario as developed does not incorporate local planning concerns and issues as articulated in local general plans, zoning and subdivision ordinances, and other local planning documents. In this sense, the Baseline Scenario is neither explicitly "pro-market" nor "pro-planning." In counties where recent development patterns have principally been a function of market factors, that reality is projected to continue. On the other hand, in counties where recent development patterns have been more constrained by formal or ad hoc policies, that reality too is projected to continue.

The process of scenario building involves four steps. The first is to calculate a future development probability for each undeveloped site. This was undertaken using the land-use change model results presented in Table 1. For purposes of calculating future job and highway accessibility, no additions to the current highway system were assumed. The second step is to specify a population growth increment to be allocated and an appropriate allocation density.

The third step is to specify a list of absolute exclusion conditions denoting which sites may *not be developed* regardless of their development probability or the level of projected population growth. Four types of sites were excluded from development under the Baseline Scenario: (i) sites in public ownership, (ii) sites currently under water, (iii) sites identified as wetlands, and (iv) sites with an average slope in excess of 20 percent. Sites upon which development *is allowed* under the Baseline Scenario include flood zone sites, farmlands of all types, sites in riparian areas, and sites presumed to be habitat to one or more threatened or endangered species. The fact that development is allowed on this latter set of sites does not mean that it is to be encouraged; rather, that under the Baseline Scenario there are no policy or planning grounds for excluding development.

The fourth and final step is to allocate prospective population growth to nonexcluded sites in order of their development probability.

4.2 Baseline Scenario Results

Statewide Baseline Scenario results indicate that projected urban development throughout the state will occur on flat sites, follow freeways, and be located in and adjacent to existing cities and urban places (Maps 1 through 4). Beyond these commonalities, growth patterns will differ significantly by region and county. Detailed land and growth projections by county, subregion, and region are provided in Appendix 1.

Map 1.

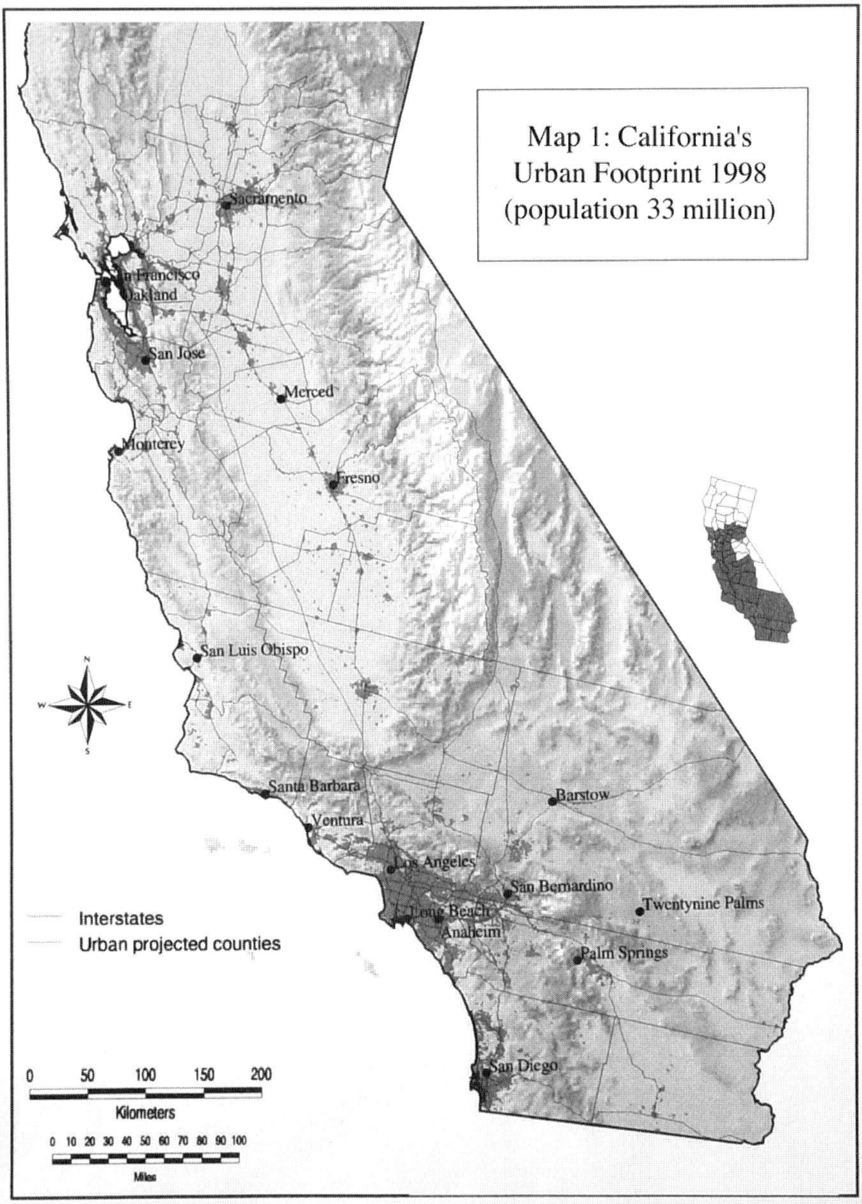

Map 1: California's Urban Footprint 1998 (population 33 million)
Sacramento
San Francisco
Oakland
San Jose
Merced
Monterey
Fresno
San Luis Obispo
Santa Barbara
Ventura
Los Angeles
Barstow
San Bernardino
Twentynine Palms
Long Beach
Anaheim
Palm Springs
San Diego
Interstates
Urban projected counties
0 50 100 150 200
Kilometers
0 10 20 30 40 50 60 70 80 90 100
Miles

Map 2.

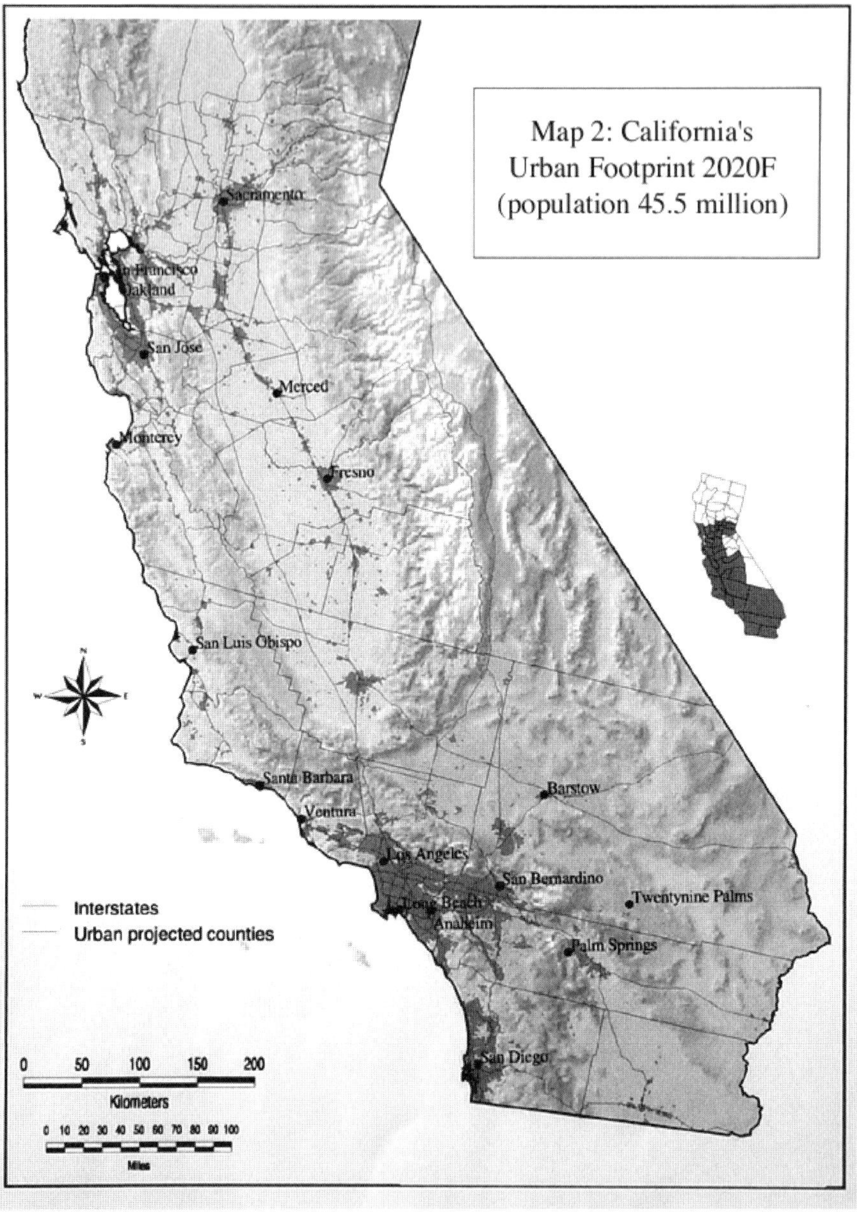
Map 2: California's
Urban Footprint 2020F
(population 45.5 million)
Sacramento
San Francisco
Oakland
San Jose
Merced
Monterey
Fresno
San Luis Obispo
Santa Barbara
Ventura
Los Angeles
San Bernardino
Long Beach
Anaheim
Barstow
Twentynine Palms
Palm Springs
San Diego
Interstates
Urban projected counties
0 50 100 150 200
Kilometers
0 10 20 30 40 50 60 70 80 90 100
Miles

Map 3.

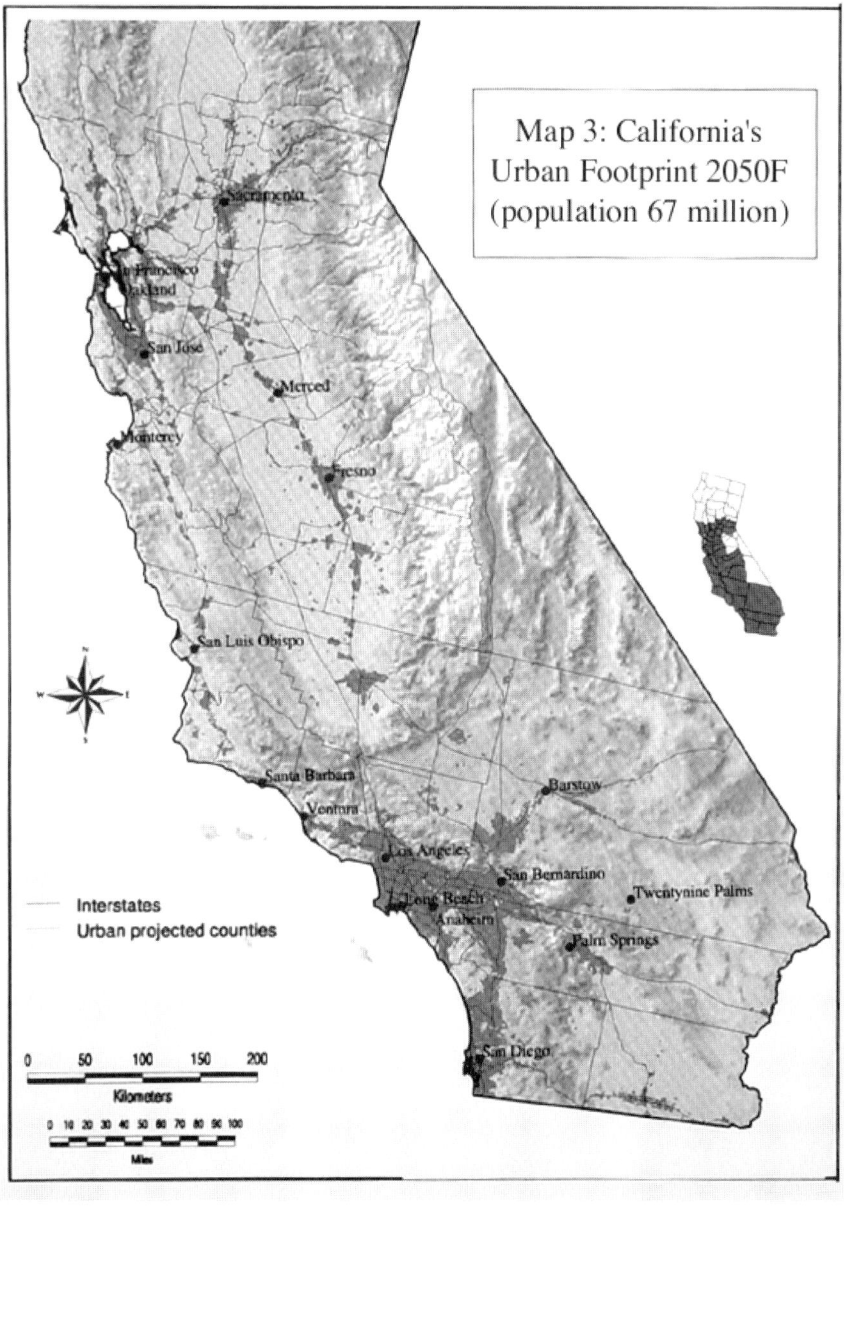
Map 3: California's
Urban Footprint 2050F
(population 67 million)
Sacramento
San Jose
Merced
Monterey
Fresno
San Luis Obispo
Santa Barbara
Ventura
Los Angeles
Barstow
San Bernardino
Twentynine Palms
Long Beach
Anaheim
Palm Springs
San Diego
Interstates
Urban projected counties
0 50 100 150 200
Kilometers
0 10 20 30 40 50 60 70 80 90 100
Miles

Map 4.

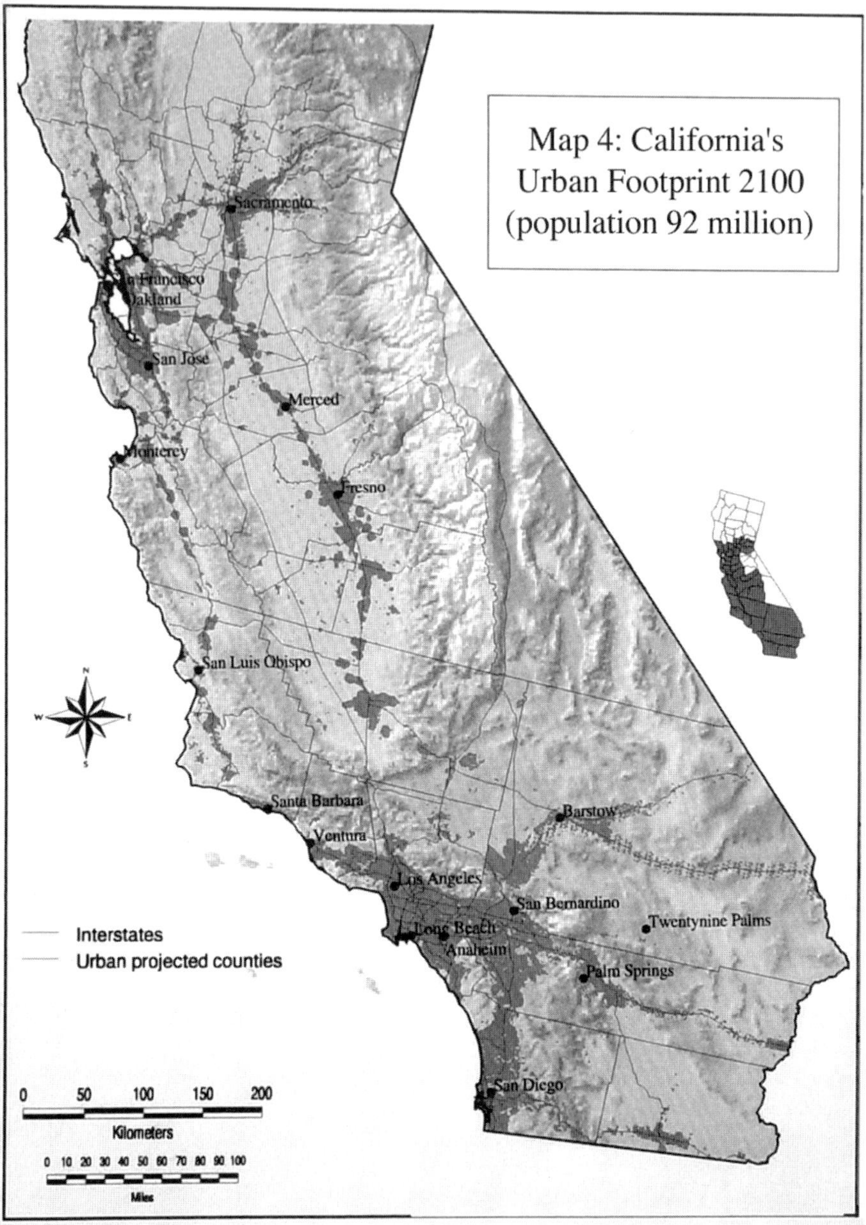
Map 4: California's
Urban Footprint 2100
(population 92 million)
Sacramento
Oakland
San Jose
Merced
Fresno
San Luis Obispo
Santa Barbara
Ventura
Los Angeles
Barstow
San Bernardino
Twentynine Palms
Anaheim
Palm Springs
San Diego
Interstates
Urban projected counties
0 50 100 150 200
Kilometers
0 10 20 30 40 50 60 70 80 90 100
Miles

Starting in the south and moving north:

Southern California: San Diego, Orange, and Imperial Counties. Urban development in the San Diego-Orange-Imperial County subregion will increase from about 245,000 hectares in 1998, to 301,000 hectares in 2020, to 385,000 hectares in 2050, to 479,000 hectares in 2100. Urban growth in the San Diego-Orange-Imperial subregion will account for about one-quarter of all new urban development in Southern California.

More than two-thirds of the region's projected urban growth will occur in San Diego County. Historically, most urban development in San Diego County has been located within ten miles of the Pacific coast. As these areas were built out in the 1980s and 1990s, growth leapfrogged north into southwestern Riverside County and, to a lesser extent, east up the foothills of the Santa Ana Mountains.

These trends will continue into the foreseeable future. If current trends continue, the I-15/Temecula area in southwestern Riverside County will be substantially built-out by about 2020, and development will have begun backfilling northern San Diego County. By 2050, Camp Pendleton, which separates San Diego and Orange counties, will be completely encircled by urban development. By 2100, if current trends continue, northern San Diego County and southwestern Riverside County will be completely urbanized, and intense urban growth will have moved eastward along Interstate-10 into central San Diego County.

Most of Orange County's projected population growth will take the form of high-density infill. Thus, while Orange County will account for a significant share of the Southern California's population growth by 2100, it will account for a far lesser share of the region's projected urban expansion—on the order of only two to four percent, depending on the period. By 2050, almost all undeveloped lands in Orange County west and north of the San Bernardino Mountains will have been developed.

The situation is the opposite for Imperial County, which will account for only two percent of the region's population growth between 1998 and 2100, but about five percent of the increase in its urban land area. All of Imperial County's projected urban growth between 1998 and 2100 will occur along Interstate-8, most of it within ten miles of El Centro.

Southern California: Los Angeles County. Except for a few areas in the San Fernando Valley, Los Angeles County is almost entirely built-out southwest of the San Gabriel Mountains. As a result, most of Los Angeles County's projected population growth during the 21st century will take the form of infill and redevelopment. Currently, Los Angeles County's urban and suburban footprint occupies about 307,000 hectares of land. By 2100, it will have grown to 361,000 hectares. Thus, while Los Angeles County will account for 30 percent of Southern California' population growth during the 21st century, its share of the region's urbanized land area growth will be just under six percent. Spatially, Los Angeles County will continue its inexorable push northward and eastward, filling out all of eastern Los Angeles County by 2020, and most of the Santa Clarita Valley by 2050.

Southern California: Ventura and Santa Barbara Counties. Urban development in Ventura and Santa Barbara counties will increase from about 63,000 hectares in 1998, to 78,000 hectares in 2020, to 113,000 hectares in 2050, to just under 150,000 hectares in 2100. Depending on the time period, growth in the Ventura and Santa Barbara subregion will account for 7–10 percent of new urban development in Southern California during the 21st century.

Being closer to Los Angeles, Ventura County will grow more and earlier than Santa Barbara county. Ventura County's urban footprint will expand by 11,000 hectares between 1998 and 2020, 17,000 hectares between 2020 and 2050, and 18,000 hectares between 2050 and 2100. These correspond to percentage increases of 28, 35, and 27 percent, respectively. Spatially, Ventura County will continue growing in northwestern direction. The Highway 101 corridor from Calabassas to Ventura and the Highway 118 corridor from Simi Valley to Moorpark will both be built-out by 2020. By 2050, development along the Highway 101 and 118 corridors will have merged, creating a continuous 20-mile westward extension of the San Fernando Valley.

Santa Barbara County should be able to continue resisting Southern California's extreme growth pressures for about another 20 years, increasing its urban footprint by only 4,000 hectares. After 2020, development activity should pick-up: between 2020 and 2100, Santa Barbara's urban footprint should increase by over 200 percent. With growth in southeastern Santa Barbara County limited by the Santa Ynez Mountains, most new development will occur along the Highway 101 corridor north from Buellton to Santa Maria. Indeed, by 2100, the entire Highway 101 corridor from downtown Los Angeles north to Santa Maria will be essentially built-out.

Southern California: The Inland Empire. With coastal areas running out of buildable land, the real development action in Southern California during the 21st century will be in Riverside and San Bernardino Counties—the Inland Empire. Urban development in the Inland Empire will increase from 208,000 hectares in 1998 to just over 800,000 hectares in 2100, an increase of nearly 400 percent. Sixty percent of new urban development in Southern California during this century will occur in the Inland Empire.

San Bernardino and Riverside counties will grow at about the same rate. Riverside County's urban footprint will increase in size from about 100,000 hectares in 1998 to just under 400,000 hectares by 2100. San Bernardino County's urban footprint will grow from 110,000 hectares in 1998 to 411,000 hectares by 2100.

The two counties' growth patterns will also be similar: development will proceed west to east, along Interstates 10 and 215 in Riverside County and along Interstates 10, 15, and 40 in San Bernardino County. By 2020, almost all remaining developable lands within a ten-mile radius of the Ontario Airport—the current growth center of the Inland Empire—will be built-out, whether in San Bernardino or Riverside counties. Development will continue apace in the Victorville-Apple Valley-Hesperia area of San Bernardino County and the Perris-Hemet-Moreno Valley area of Riverside County. By 2050, both areas will have emerged as major metropolitan centers, and the Inland Empire will be entirely built-out west of the line connecting Hemet in Riverside County and Yucaipa in San Bernardino County. North of the Cajon Pass, intense suburban development will reach the Barstow area and points east along Interstates-5 and -40 by 2030. By 2050, the Coachella Valley (stretching from Palm Springs to Indio) will be built-out south of Interstate-10; by 2100, the north side of the Coachella Valley will have been developed.

The Southern San Joaquin Valley: Kern, Kings, Fresno, Madera and Tulare Counties. Urban development in the southern end of the San Joaquin Valley will grow from 118,000 hectares in 1998 to 418,000 hectares in 2100—an increase of nearly 250 percent. Nearly three-quarters of new urban development in the entire San Joaquin Valley (stretching from Kern County in the south to San Joaquin County in the north) will be in the southern subregion. As it has since the turn of the 20th century, new development in the southern San Joaquin Valley will be concentrated along the Highway 99 corridor—with or without the construction of a high-speed rail system. For the most part, development will occur from the north to the south, connecting Fresno and Visalia by 2030 and extending south to Tulare and Corcoran by 2050. By 2100, the entire corridor will be urbanized, and active farmlands will have been pushed to the east and west. New development will also follow Highway 99 south of Bakersfield toward Los Angeles.

Almost one-third of the region's urban growth will occur in Kern County. Between 1998 and 2100, Kern County's urban footprint will expand from 65,000 hectares to nearly 160,000 hectares. Currently, almost all urban development in Kern County is concentrated in and around Bakersfield. With the city's urban footprint likely to grow three-fold by 2100, Bakersfield will continue to dominate Kern County's urban landscape. Even so, new and smaller urban nodes will also develop around the cities of Shafter and Delano by 2020, Wasco and Tehachapi by 2050, and Arvin and Mojave by 2100.

The other major locus of future urban growth in the southern San Joaquin Valley will be Fresno County. Between 1998 and 2100, Fresno County's urban footprint will expand three-fold from 38,000 hectares to 119,000 hectares. Almost all new urban development in Fresno County will occur at the outskirts of the City of Fresno or along Highway 99.

Tulare County, which lies along Highway 99 between Kern and Fresno counties, will also grow significantly more urban during the 21st century. If current trends continue, Tulare's urban footprint will grow in size from 20,000 hectares in 1998 to 76,000 hectares in 2100. While initially clustered around Visalia and Tulare, by 2100, the entire Highway 99 corridor in Tulare County will be urbanized. Madera and Kings County will also see significant, albeit somewhat less, amount of additional urban growth throughout the 21st century.

The Northern San Joaquin Valley: Kern, Kings, Fresno, Madera, and Tulare Counties. The northern San Joaquin Valley extends from Lodi and Stockton in San Joaquin County in the north; south along Highway 99 to Modesto, Ceres, and Turlock in Stanislaus County; and then further south to Livingston and Merced in

Merced County. New development in the northern San Joaquin Valley subregion will be principally fed by unaccommodated eastward overflow growth from Bay Area. Altogether, urban development in the northern San Joaquin Valley will grow from 62,000 hectares in 1998 to 180,000 hectares in 2100—an increase of nearly 200 percent. As in the southern part of the San Joaquin Valley, new development in the north will be concentrated along the Highway 99 corridor. By 2100, the entire Highway 99 corridor will be developed to a width of 10 to 20 miles in San Joaquin County, down to 5 to 10 miles in Merced County.

About half of the northern subregion's urban growth will occur in San Joaquin County. Between 1998 and 2100, San Joaquin County's urban footprint will expand from 29,000 to 85,000 hectares. By 2020, the Interstate-205 corridor connecting Tracy and Manteca will be mostly built-out. By 2050, urban development will extend continuously along the Highway 99/I-5 corridor from Lodi in the north to Ripon in the south. By 2100, San Joaquin County's urban footprint will rival Santa Clara's in size.

Stanislaus County will also grow substantially, adding 55,000 hectares of new urban development by 2100. Most of Stanislaus County's new development will occur in and around the cities of Modesto and Turlock. Further south, Merced County's urban footprint will expand from 12,000 hectares in 1998 to over 40,000 hectares by 2100. Growth in Merced County will generally proceed north to south starting in Livingston, and then moving south to Atwater, and later to Merced. One wildcard in the future development of Merced County is the new University of California-Merced campus, the presence of which was not included in the baseline model runs. If, as is intended, UC-Merced serves as an engine of economic development, urban growth in Merced County could well exceed these estimates.

Northern California: The Monterey Bay Area and San Luis Obispo County. We have included the Monterey Bay Area and San Luis Obispo County as a subregion of Northern California—something that isn't usually done, at least not yet. As the Silicon Valley continues to grow southward, its economic sphere of influence will begin to envelop Monterey, San Benito, Santa Cruz, and even the northern section of San Luis Obispo County.

Urban development in these four counties will increase continuously from 49,000 hectares in 1998 to more than 150,000 hectares by 2100. Depending on the time period, urban growth in this subregion will account for between a third and one-half of new urban development in Northern California during the 21st century. Unlike the central Bay Area, where significant population growth will occur as infill, most population growth in the Monterey Bay Area and points south will occur as "greenfield" development.

Monterey County, being more directly connected to Santa Clara County along Highway 101, will grow more and earlier than its three subregional neighbors. From its current size of 20,000 hectares, Monterey County's urban footprint will expand to 51,000 hectares by 2050 and 75,000 hectares by 2100. Within Monterey County, the wave of urban development will move north to south, enveloping

Prunedale and Salinas by 2020, Gonzales and Soledad by 2050, and reaching as far south as King City by 2100. Indeed, by 2050, the central spine of the Salinas River Valley—which includes some of the most fertile farmland in the world—will be essentially built-out.

Further to the south, San Luis Obispo County's urban footprint will also expand significantly from 15,000 hectares in 1998 to nearly 46,000 hectares by 2100. Growth will occur radially around the cities lining the Highway 101 corridor. These include Paso Robles, Atascadero, San Luis Obispo, and Pismo Beach.

Because they lack flat, accessible, and easily-serviced raw land, Santa Cruz County and San Benito County will grow more moderately—at least in comparison to Monterey County and San Luis Obispo County. If current trends continue, Santa Cruz County's urban footprint will expand from 12,000 hectares in 1998 to over 20,000 hectares in 2100. Most of this growth will occur in the Watsonville area. To the east, San Benito County's urban footprint will grow from about 3,000 hectares in 1998 to 10,000 hectares in 2100.

Northern California: The Central San Francisco Bay Area. Urban development in the central San Francisco Bay Area—encompassing Alameda, Contra Costa, San Francisco, San Mateo, and Santa Clara counties—currently occupies a 220,000 hectare footprint. With so little undeveloped land remaining adjacent to the San Francisco Bay, most new development in this subregion will occur east of the Oakland/East Bay Hills and south of San Jose. If current trends continue, the Central Bay Area's urban footprint will grow in size to 240,000 hectares by 2020, 260,000 hectares by 2050, and 284,000 hectares by 2100. This is a relatively modest level of growth compared to California's other urban areas and reflects the fact that most of the subregion's population growth will take the form of infill and redevelopment. Although it is currently home to more than five million residents and more than 75 percent of the Northern California region's population, over the next century the Central Bay Area will account for less than 30 percent of the region's projected urban growth.

Already mostly urbanized, Santa Clara County has little flat and accessible land available for future development and most of what it does have is in the central and southern part of the county. As a result, Santa Clara County's urban footprint will grow from its current size of 73,000 hectares to 91,000 hectares by 2100. Almost all of this increase will occur within the Highway 101 corridor south of San Jose.

As is the case today, many of those who work in Santa Clara County will live in an adjacent county. Neighboring Alameda County's urban footprint, for example, will expand from its current size of 57,000 hectares to 79,053 hectares by 2100. Most this increase will occur in and around three cities: Dublin, Pleasanton, and Livermore. Further north, Contra Costa County will also experience significant development pressure as its urban footprint grows from its current size of 56,000 hectares to 71,000 hectares in 2100. Contra Costa's new urban growth will be divided between the Interstate-680 corridor connecting Martinez and Pleasanton, and the Highway 4 corridor connecting Concord and Brentwood. Development will also be continuously climbing the foothills of Mt. Diablo and the western side of the East Bay Hills. Should a major highway be built on the eastern side of Mt. Diablo—something we have not included in our projections—Mt. Diablo would soon be completely encircled by urban growth.

Over on the San Mateo Peninsula, San Francisco is already entirely built-out and will accommodate all its projected population growth through infill and redevelopment. San Mateo County will also grow principally via infill and redevelopment; between 1998 and 2100, its urban footprint will expand by less than 6,000 hectares. Most of this growth will occur adjacent to the San Francisco Bay or south of Pacifica along Highway 1.

Northern California: The North Bay Area. The north San Francisco Bay Area includes Marin, Sonoma, Napa, and Solano counties. Currently, urban development in this region is organized into a series of separate suburban valleys along Highway 101 in Marin and Sonoma counties, Highways 29 and 12 in Napa County, and Interstate-80 in Solano County. As of 1998, the North Bay subregion included 15 percent of the Northern California Region's population and at 73,000 hectares, about 21 percent of its urbanized area. By 2100, should present trends continue, the urbanized area of the North Bay subregion will have increased to 133,000 hectares.

Most of this increase will take place in Sonoma and Solano counties. Sonoma County's urban footprint will likely grow from its current size of 27,000 hectares to nearly 55,000 hectares by 2100. Almost all of this increase will occur within five miles of Highway 101. Indeed, by 2050, the Highway 101 corridor will be continuously developed from the Sonoma-Marin county line north through the city of Healdsburg.

A similar corridor-centric form will characterize urban growth patterns in suburban Solano County: almost all of Solano County's growth will occur within 10 kilometers of Interstate 80. Altogether, Solano County's urban footprint will grow from its current size of 22,000 hectares to more than 44,000 hectares by 2100.

Compared to Sonoma and Solano counties, Marin and Napa counties will hardly grow at all—although their relative growth will seem sizeable. Marin County's urban footprint will expand from its current size of just over 16,000 hectares to 19,000 hectares by 2100. Most of Marin's projected new development will occur along the Highway 101 corridor in and around Novato. Lacking good freeway access, Napa County will also experience only moderate growth, adding about 6,000 hectares of new urban development by 2100.

The Sacramento Region: Sacramento County: The Sacramento Region extends from Yolo County in the west to Lake Tahoe in the east, from Sutter and Yuba counties in the north, to Isleton (Sacramento County) in the south. By 2100, if current trends continue, the region's urban footprint will have more than doubled from its current size of 112,000 hectares to nearly 225,000 hectares.

At 61,000 hectares, Sacramento County alone accounts for just over half of the region's total urban footprint. Located at the confluence of the Sacramento and American Rivers, Sacramento County still has ample flat land upon which to grow—mostly to the south and east—and by 2100, its urban footprint will likely exceed 100,000 hectares. Highway 50, which currently forms the southern boundary of development in eastern Sacramento County, will be likely be breached by 2025 as urban growth continues pushing eastward. New growth will also extend northward along Interstate 5 and California Highway 70. All told, Sacramento County will account for about one-third of the Sacramento Region's growth during the 21st century.

The Sacramento Region: Foothill Subregion. The Foothill subregion consists of the western sides of El Dorado, Placer, and Nevada counties. Urban growth in these counties generally takes two or three forms. On the west side, adjacent to Sacramento County, urban development takes a mostly suburban form, consisting of large, moderate-density single-family subdivisions, retail strip and power centers, and the occasional free-standing office building. Ten miles to the east, along Interstate 80 and Highway 50 in and around the hills of Auburn and Placerville, growth consists of smaller residential developments of larger lot sizes, sprinkled at the edges of existing cities and towns. Further off the beaten track abutting local roads, new development typically takes the form of clusters of large houses on large lots, some with fenced-in grazing and farmlands. Known as "ranchettes," these developments occupy significant land areas but accommodate relatively few residents.

Currently, the urban footprint of the Foothill subregion, excluding ranchettes, exurban, and vacation development, total about 32,000 hectares By 2050, urbanization in the Foothill subregion will have nearly doubled; by 2100, it will be approaching 80,000 hectares. Most of the subregion's growth will occur in Placer County along Interstate 80. Indeed, by 2100, the I-80 corridor will be completely built out to a width of five kilometers from Roseville, past Auburn to Meadow Vista. Lesser although still sizeable amounts of development are projected for El Dorado and Nevada counties. Growth in El Dorado County will be focused along Highway 50 in and around the Placerville area. In Nevada County, new urban development will favor the Grass Valley-Nevada City area. All together, the three-county Foothills subregion will account for about 40 percent of the growth of the Sacramento Region during the 21st century.

The Sacramento Region: Yolo County. Located at the western end of the Sacramento Region, Yolo is separated from Sacramento County and the rest of the region by the Sacramento River. Urban growth in Yolo County has long followed a city-centric pattern—the result of the county's long-standing commitment to conserving farmlands and discouraging sprawl. As of 1998, Yolo County's urban footprint occupied just over 10,000 hectares of land. By 2100, this total is projected to double to just over 21,000 hectares. Unless they act to limit growth—something the City of Davis is periodically wont to do—just about all of this increase will occur in and around the three cities that line Interstate 80: Dixon, Davis, and West Sacramento. Woodland and Winters, Yolo's two other significant cities, will grow less dramatically. Altogether, Yolo County will account for about ten percent of the Sacramento Region's urban growth in the 21st century.

The Sacramento Region: Sutter and Yuba Counties. Yuba and Sutter counties are the least populous, least urbanized counties in the Sacramento Region. At just under 9,000 hectares, urban development in Yuba and Sutter counties currently account for only percent of the region's total. If current trends continue, Yuba and Sutter's urban footprint will swell to 17,000 hectares by 2050, and 23,000 hectares by 2100. These are big increases by the standards of Yuba and Sutter counties. They are less large when compared to the region a whole: altogether, Yuba and Sutter counties will account for about 12 percent of the region's urban growth during the 21st century. Spatially, most of this subregion's growth will occur in southeastern Sutter County near the Sacramento County border.

5.0 Impact of Baseline Scenario on Landscape Change

The conversion of undeveloped land to urban uses generates three types of effects on the landscape: (i) it reduces the amount of undeveloped land still available; (ii) it alters the patch size, shape, and fragmentation level of the remaining undeveloped landscape; and, (iii) it alters both the amount and quality of the resource and environmental services provided by undeveloped lands.

This section undertakes to measure effect (i) resulting from the Baseline Scenario for the periods 1997–2020, 2020–2050, and 2050–2100. This effect can be measured using many of the same digital data layers used to derive the Baseline Scenario. It is important to note up front that measuring landscape change is not the same thing as valuing landscape change. Valuing landscape change requires incorporating human preferences regarding relative scarcity, accessibility, existence value, and a whole host of other attributes. Valuation can be undertaken through an analysis of market and nonmarket transactions or through the use of survey research methods. Neither method is employed here.

5.1 Hillsides and Steeply Sloped Land

Statewide, projected urban growth presents a relatively small threat to steep hillsides. Among the 45 counties for which we developed detailed urban growth projections, we project that an additional 8,200 hectares of steeply sloped land—that is, land with a slope in excess of 15 percent—will be developed by 2020. By 2050 and 2100, we project that urban growth will have consumed an additional 38,000 and 55,000 hectares of steeply sloped land. These growth increments account for only .1, .4, and .6 percent of the current hillside land area of these counties.

The counties projected to suffer the largest absolute hillside losses by 2050 and 2100 are all in Southern California. They are San Diego County (14,600 hectares or four percent of the county's remaining steep hillsides), Riverside County (13,900 hectares or three percent), Los Angeles County (8,800 hectares or three percent and San Bernardino County (6,300 hectares or one percent). The only non–Southern California counties projected to suffer significant hillside losses due to urbanization are Placer and El Dorado. Because it is extremely flat, Sacramento County is likely to suffer minimal absolute hillside losses (12 hectares) but significant relative losses (12 percent).

Readers should remember that these projections assume a continuation of current development trends and patterns. Should these patterns shift in ways that make hillside development easier from a regulatory perspective, less costly from a development perspective, or more attractive from a market perspective, it is quite conceivable that amounts of hillside loss could be much greater, particularly in counties like San Diego, Ventura, Orange, Santa Barbara, Santa Clara, and Marin, which are all running out of accessible flat land near urban centers.

5.2 Wetlands

Principally for planning and regulatory reasons, wetland development is growing increasingly difficult throughout California. Counties with large amounts of wetlands currently in agricultural use. These counties are looking for ways to keep them that way. Counties with few remaining wetland areas are vigorously trying to protect and enhance them.

Statewide, projected urban growth presents a small but significant threat to wetlands, particularly those identified as part of the National Wetland Inventory.[7] Among the 28 California counties with significant remaining wetlands, which are threatened by urban growth, we project that an additional 12,000 hectares of wetlands will be developed by 2020. By 2050 and 2100, respectively, we project that urban growth will have likely consumed an additional 26,000 and 42,000 acres of wetlands. In percentage terms, these growth increments account for only one, two, and three percent of the current wetlands inventory.

The counties projected to suffer the largest absolute wetlands losses by 2050 and 2100 are mostly in the northern San Joaquin Valley, the southern Sacramento River Valley, or adjacent to the San Francisco Bay. They include San Joaquin County (8,600 hectares or 11 percent of the county's remaining wetlands), Sutter County (4,300 hectares or seven percent), Sonoma County (3,200 hectares or 26 percent), Solano County (2,500 hectares or five percent), and Alameda County (2,300 hectares or 33 percent). A number of additional counties are facing moderate absolute wetland losses due to urbanization but large percentage losses: Marin County (1,700 hectares or 24 percent), San Mateo County (1,300 hectares or 42 percent), San Diego County (1,200 hectares or 15 percent), and Santa Clara (700 hectares or 15 percent). At the other extreme, Sacramento, Merced, and Yolo counties are all facing moderate absolute wetland losses but small relative losses.

Issues of wetland conservation and protection go well beyond consideration of absolute and percentage losses. Wetlands are typically prime habitat for a wide variety of plant and animal species; many of which are the national threatened and endangered list. Wetlands also play an important role in insuring the health of adjacent habitat areas and in flood control. Because wetlands are not interchangeable, questions of how and where projected urbanization is likely to affect wetland quality may dominate questions of absolute loss.

5.3 Riparian Areas

Riparian zones are the land areas around rivers, streams, lakes, and permanent wetlands. They are typically but not exclusively characterized by woody, fast-growing vegetation and by water-oriented bird, animal, and insect species. Inventories of riparian areas have thus far been developed for the San Francisco Bay and San Joaquin Valley but not for the rest of the state. To augment these more limited data sources, we generated a statewide, 100-meter riparian zone data layer by buffering all inland rivers, streams, and lakes listed in the 2000 Census TIGER file. Although comprehensive and consistent, this method tends to over-estimate the total amount of riparian area while underestimating the area of specific riparian zones.

The counties projected to suffer the largest absolute riparian losses by 2050 and 2100 are mostly in southern and coastal California. They include San Bernardino County (52,200 hectares or six percent of the county's remaining riparian zone land area), Riverside County (51,000 hectares or 13 percent), San Diego County (22,000 hectares or 11 percent), Imperial County (14,000 hectares or six percent), and Kern County (14,000 hectares or three percent). A number of additional counties are facing moderate-to-small absolute riparian zone losses due to urbanization but large percentage losses: Stanislaus County (5,000 hectares by 2100 or 12 percent), San Joaquin County (4,900 hectares or 12 percent), Sacramento County (4,600 hectares or 12 percent), Alameda County (3,800 hectares or 13 percent), Orange County (2,700 hectares or 11 percent), and Santa Cruz County (2,400 hectares or 10 percent). San Diego County (1,200 hectares or 15 percent), and Santa Clara (700 hectares or 15 percent).

As is the case for wetlands, riparian zone quality varies widely. Some provide rich habitats for a diverse variety of plant and animal species, others are ecologically narrower. Most of the state's remaining riparian lands south of Sacramento and west of the Sierras border urban development, active farmlands, or grazing lands and, as a result, have suffered severe degradation. Thus, in many areas, riparian zone restoration is as important as riparian zone protection.

5.4 Prime Farmlands

The California Farmland Mapping and Monitoring Project (CFMMP) biannually collects detailed spatial data regarding the status of different types of Farmland in 47 California counties, including all urban counties except San Francisco. Prime farmland is defined by the CFMMP as land used for the production of irrigated crops at some time during the previous four years and which meets the highest soil moisture, pH, erodability and permeability, and soil rooting depth criteria. At the time of this work, the most recent year for which complete data were available was 1998.

Projected urban growth presents a significant threat to the state's remaining supplies of prime farmland, especially in the San Joaquin, Monterey, and Imperial Valleys. Among the 35 California counties in which prime farmlands are threatened by urban growth, we project that 52,000 hectares of prime farmland will have been converted to urban uses by 2020. By 2050 and 2100, respectively, we project that urban growth will have likely consumed an additional 165,000 and 297,000 hectares of prime farmlands. In percentage terms, these growth increments account for 3, 9, and 17 percent of the current inventory of prime farmlands.

Most at risk from urban growth are prime farmlands in the San Joaquin and Monterey valleys. Assuming present trends continue, Fresno County will lose 51,600 hectares of prime farmland by 2100, a drop of 35 percent. Nearby Kern County will lose 42,100 hectare of prime farmland to projected urban growth, a 19 percent decline. San Joaquin, Monterey, and Riverside Counties will each lose in excess of 20,000 hectares. Five counties will lose more than half of the precious little prime farmland they still have: San Bernardino (78 percent), San Diego County (58 percent), Orange County (57 percent), Alameda County (51 percent), and Santa Clara County (50 percent). For counties in Southern California, these losses will be immediate—that is, they will mostly occur by 2020. Among Bay Area and coastal counties they will occur over a longer period. Among San Joaquin Valley counties, prime farmland losses will be continuous throughout the century.

The situation may not be quite as bleak as these numbers would make it seem If current trends continue, even as large amounts of prime farmland are lost to urban growth, farmers will likely be "developing" new prime farmlands in other locations, mostly by extending irrigation to grazing and secondary farmlands. While not of the soil quality of the prime farmlands being lost, assuming sufficient water is available at the right price, these new farmlands should easily sustain California's agricultural economy. Potential opportunities for new prime farmland development are most plentiful in the San Joaquin and Monterey valleys (where urbanization will pose less of a threat to grazing lands) but extremely limited in Southern California and the Bay Area. Thus, for purposes of protecting prime farmland from incursion by urban growth throughout the state, policymakers and farmland preservationists should perhaps concentrate their efforts in Riverside, Imperial, San Bernardino, Ventura, San Diego, Alameda, and Contra Costa counties, among others.

5.5 State and Locally Important Farmlands

Farmlands of state importance are those similar to prime farmland but with minor shortcomings, such as greater slopes or less ability to store soil moisture. *Farmland of local importance* is determined by each county's board of supervisors and a local advisory committee. For purposes of tracking farmland loss due to projected urbanization, we have grouped state and locally-important farmland.

Projected urban growth presents a significant threat to the state's remaining supplies of state and locally important (S&LI) farmlands, especially in Riverside, Imperial, San Diego, Sacramento, and San Joaquin counties. Among the 36 California counties in which S&LI farmlands are threatened by urban growth, we project that an additional 74,000 hectares of S&LI farmland will have been converted to urban uses by 2020. By 2050 and 2100, respectively, we project that urban growth will have likely consumed an additional 173,000 and 268,000 hectares of S&LI farmland. In percentage terms, these growth increments account for only 4, 9, and 14 percent of the current inventory of S&LI farmlands.

Most at risk from urban growth are S&LI farmlands in the Inland Empire and in the central Sacramento-San Joaquin Valley. Assuming present trends continue, Riverside County will lose 66,000 hectares of S&LI farmland by 2100, a drop of 54 percent. Nearby Imperial County will lose 25,000 hectares to projected urban growth, an 18 percent decline. San Diego and Sacramento counties each lose in excess of 20,000 hectares, and San Joaquin will lose nearly that much. In addition to Riverside, five counties will lose more than half of their S&LI farmlands by 2100: Alameda County (72 percent), San Bernardino County (62 percent), Monterey County (51 percent), San Diego County (51 percent), and Sacramento County (50 percent). While Bay Area and coastal counties will suffer significant relative losses, except for Monterey and Sonoma counties, their absolute losses will not be that great. In terms of timing, most losses will be continuous throughout the forecast period. Whether or not these losses will be offset through the irrigation and conversion of grazing land will depend on many factors: water availability, commodity prices, labor costs, and changing environmental regulations.

5.6 Unique Farmlands

The CFMMP identifies *unique farmland* as land of lesser-quality soils used for the production of the state's leading agricultural crops. Unique farmlands are typically irrigated but may also include nonirrigated orchards or vineyards as found in some climatic zones in California. Despite their lower soil quality, many of California's highest value crops are grown on unique farmlands. Unique farmlands must have been cropped at some time during the four years prior to the mapping date.

Projected urban growth presents a significant threat to the state's remaining supplies of unique farmlands, especially in San Diego and Riverside counties. Among the 36 California counties in which unique farmlands are threatened by urban growth, we project that an additional 13,000 hectares of unique farmland will have been converted to urban uses by 2020. By 2050 and 2100, respectively, we project that urban growth will have likely consumed an additional 47,000 and 77,000 hectares of unique farmlands. In percentage terms, these growth increments account for only two percent, five percent, and eight percent of the current inventory of unique farmlands in the counties studied.

Most at risk from urban growth are unique farmlands in and around the Inland Empire and in the central San Joaquin Valley. Assuming present trends continue, San Diego County will lose 20,000 hectares of unique farmlands by 2100 or three-quarters of its current stock. Nearby Riverside will lose 10,000 hectares to projected urban growth, a 59 percent decline. Riverside will, in fact, lose more than half of their unique farmlands to urban development by 2100. In the central San Joaquin Valley, Fresno and Madera counties will each lose in excess of 5,000 hectares. Nearby Merced County will lose 3,800 hectares.

In addition to San Diego and Riverside counties, San Bernardino County (86 percent), Alameda County (62 percent), and Orange County (62 percent) would be the other big losers of unique farmlands. Indeed, because unique farmlands are often the left over from cities, hillsides, and prime farmlands, all of California's coastal counties south of San Francisco and most of its Central Valley counties are facing significant losses of unique farmlands due to projected urban growth. Presumably, as in the case of prime farmlands and S&LI farmlands, it may be possible to offset some of these losses through irrigation of otherwise fertile lands currently in grazing use. The specific potential for such conversion is unknown.

5.7 Grazing Lands

The CFMMP identifies *grazing lands* as those on which the existing vegetation is suited to the grazing of livestock. This category is used only in California and was developed in cooperation with the California Cattlemen's Association, University of California Cooperative Extension, and other groups interested in the extent of grazing activities.

Projected urban growth presents a significant threat to grazing lands in Riverside, Placer, San Diego, and San Bernardino counties; a moderate threat in Orange, Ventura, Alameda, Solano, Sacramento, Los Angeles, Santa Cruz, and Santa Barbara counties; and a minor threat elsewhere in the state. Among the 35 California counties in which grazing lands are threatened by urban growth, we project that an additional 77,600 hectares of grazing land will have been converted to urban uses by 2020. By 2050 and 2100, respectively, we project that urban growth will have likely consumed an additional 216,300 and 329,600 hectares of grazing lands. In percentage terms, these growth increments account for only one, two, and three percent of the current inventory of grazing land in the counties studied.

Most at risk from urban growth are grazing lands in and around the Inland Empire. Assuming present trends continue, San Bernardino County will lose 132,400 hectares of grazing land by 2100 or about one-third its current supply. Nearby Riverside and San Diego counties will lose 28,400 and 23,500 hectares respectively—declines of 52 percent and 41 percent. Six other counties will lose more than 10,000 hectares of grazing land to projected urban growth by 2100: Kern

(22,600), Santa Barbara (16,400), San Luis Obispo (11,900), Ventura (11,600), Alameda (11,400), and Monterey (10,300).

The real threat to grazing lands may be well bigger than these numbers suggest. As urbanization consumes prime and unique quality farmlands in the San Joaquin, Monterey, and Imperial valleys, agricultural businesses may well attempt to convert grazing lands to cultivated use, mostly through the extension of irrigation. The potential for this "domino effect" to occur will depend on many factors, including land, water, and agricultural product prices and the costs of extending key infrastructure.

The trend toward ranchette and resort/vacation home development also threatens California's grazing lands. Ranchettes and rural subdivisions are typically developed at densities well below the threshold used by the CFMMP to identify urban and suburban development. As a result, it mostly goes uncounted during the CFMMP's biannual farmland inventories. In the absence of hard data, anecdotal evidence suggests that such activity is on the rise and that it is mostly occurring on grazing and ranchland. To the extent that California's urban areas continue to spin-off ranchettes and rural subdivisions, the potential threat of population growth to grazing land may be the greatest of all. Thus, while the geography of cultivated farmland in California in 2100 could very well be similar to that of today, the geography of California's grazing lands will almost certainly be different.

6.0 Key Assumptions, Caveats, and Concluding Comments

Numerous assumptions are embedded in this procedure and its components. Perhaps the most questionable is whether it is within the realm of human capability to accurately extrapolate current population and employment growth trends and urban settlement patterns far into the future—in this case, through the year 2100—and particularly in a state as changeable as California. If history teaches us anything, it is that the future is *always different* than we anticipate it will be, no matter how sophisticated our reasoning or projection techniques. For this reason, the projections developed here and in later efforts are best viewed not as forecasts, per se, but as scenarios—that is, as a set of illustrative futures designed to indicate how particular growth trends and development dynamics might play out upon California's diverse landscapes. Beyond this general caveat, there are five specific assumptions driving this analysis:

1. *The same factors that shaped land development patterns in the recent past will continue to do so in the future and in the same ways.* As previously discussed, this procedure allocates future development to individual sites based on their projected development probability. These probabilities are estimated using the results of a statistical model calibrated for the period 1988–98. While the exact role of particular factors varies by region, several influences are consistently important. These include proximity to freeways, access to jobs, site slope, and

site incorporation status. Other factors such as farmland and wetland status vary more widely in their importance. To the extent that these factors are less important in the future, are important in different ways, or as is even more likely, that other factors become important, the model results will vary widely than what is presented here.

2. *Jobs will continue decentralizing within California's four major urban regions—Southern California, the greater San Francisco Bay Area, the Sacramento region, and the southern San Joaquin Valley.* Taking advantage of improved freeway access, less expensive land, and lower development costs, job growth during the last 50 years has favored suburban locations over core cities. To the extent that this trend continues—given the increasing importance of telecommunications in shaping economic geography and in the absence of countervailing policies, there is no reason to believe that it should not—decentralizing job growth will continue to pull population outward, leading to more decentralized growth patterns.
3. *California's population will continue to grow and at more or less the same rate and in the same spatial pattern as projected by the California Department of Finance.* For consistency's sake, we rely on county population projections developed by the California Department of Finance through 2040. (DoF population projections are calculated by extrapolating current fertility and migration trends.) Thereafter, we extrapolate and trend downward the annualized county growth rates embedded in the DoF population projections. This approach yields a statewide population of 68 million in 2050 and 92 million in 2100.

   As large as these numbers are, they are hardly inconceivable. Since 1940, thanks to its robust economy, benign weather, and location on the Pacific Rim, California has been adding population at a steady rate of about five million persons every decade. Should this trend continue, California's 2100 population would exceed 85 million. On the flip-side, should California's economy falter or the state's high cost of living start to choke off further job growth, the state's population could easily plateau around 50 million and, though it seems unthinkable today, perhaps even trend downward.
4. *Average infill rates and population densities will increase with additional development.* It is an axiom of economics that scarce resources are used more intensely than plentiful ones. Following this logic, as available supplies of developable land are used up, developers seek ways to use remaining land more intensely, either by increasing densities or through redevelopment. Thus, both development densities and infill activity should increase with population growth. Counteracting this tendency is the desire of many residents to preserve a rural or suburban lifestyle. Thus, there are many parts of California where infill activity and development densities are below what theory suggests they should be. For the purposes of constructing a baseline scenario, we assume that future infill activity and development densities will follow the upward trend lines reported in Figures 4 and 5. To the extent that it does not, additional greenfield sites will be needed to accommodate projected population growth.

5. *With respect to the Baseline Scenario, no new freeways or intra- and inter-regional rapid transit systems will be developed. Freeway road travel speeds will remain at current levels.* This is perhaps the least realistic assumption of all. It is abundantly clear that California's growing population will need additional transportation infrastructure. What is unclear is what the infrastructure should be, where it should go, and how it should be planned and financed. Lacking these specifics, and for the purposes of constructing a baseline scenario, we assumed no change in transportation technology or facilities beyond what is currently available. The effect of this assumption is to direct additional growth to locations already served by transportation infrastructure rather than to new or different areas.

The exercise of projecting urban growth and change in California for the 21st century allows us to appreciate the critical relationship between the choices we make about how we grow and its impact on the environment. As the analysis shows, the expected huge population growth will more than double the present urban footprint in California *if current patterns of development are continued.* The baseline scenario is in fact a wake-up call to devise new strategies for future development. We will be testing several other scenarios with the help of this model in which different patterns of development are analyzed to assess their impact on growth. Our previous studies have shown that modest changes in density of development and more efficient use of infill sites can significantly reduce the need for greenfield development (Sandoval and Landis 2000; Landis 2001). Therefore it is now essential to explore new innovative approaches to development and smart planning so that we can accommodate the expected population growth without sacrificing environmental concerns.

Appendix 1.

Appendix 1. Summary of Urban Land Area Growth Projections by County, Subregion, and Region: 1998, 2020, 2050, 2100

Major Region	County	Subregion	Urbanized Land Area (Hectares)				County Share of Regional Change		
			1998	2020F	2050F	2100F	1998–2020	2020–2050	2050–2100
Southern California	Los Angeles	Central	307,205	318,174	342,037	360,808	5.3%	6.6%	4.7%
	Imperial	South	9,682	19,834	38,365	59,615	4.9%	5.1%	5.3%
	Orange	South	109,364	116,424	122,459	129,443	3.4%	1.7%	1.8%
	San Diego	South	125,883	164,271	224,118	290,171	18.5%	16.6%	16.6%
	Subregional Total		**244,929**	**300,529**	**384,942**	**479,229**	**26.8%**	**23.4%**	**23.7%**
	Riverside	Inland Empire	97,760	162,938	270,893	389,620	31.4%	29.9%	29.8%
	San Bernardino	Inland Empire	110,329	171,155	281,363	411,287	29.3%	30.5%	32.7%
	Subregional Total		**208,089**	**334,093**	**552,256**	**800,907**	**60.7%**	**60.4%**	**62.5%**
	Santa Barbara	North	24,061	28,142	45,317	63,227	2.0%	4.8%	4.5%
	Ventura	North	39,135	50,043	67,330	85,631	5.3%	4.8%	4.6%
	Subregional Total		**63,196**	**78,185**	**112,647**	**148,858**	**7.2%**	**9.5%**	**9.1%**
	REGIONAL TOTAL		**823,419**	**1,030,981**	**1,391,882**	**1,789,802**	**100.0%**	**100.0%**	**100.0%**
Northern California	Alameda	Central	56,562	63,453	70,471	79,053	12.9%	8.5%	9.7%
	Contra Costa	Central	55,547	60,250	65,067	70,751	8.8%	5.8%	6.4%
	San Francisco	Central	9,386	9,386	9,386	9,386	0.0%	0.0%	0.0%
	San Mateo	Central	28,473	29,769	31,682	34,300	2.4%	2.3%	2.9%
	Santa Clara	Central	72,717	77,510	83,628	91,392	9.0%	7.4%	8.7%
	Subregional Total		**222,685**	**240,368**	**260,234**	**284,882**	**33.2%**	**24.0%**	**27.7%**
	Marin	North	16,073	16,590	17,718	19,373	1.0%	1.4%	1.9%
	Napa	North	8,313	9,861	11,924	14,411	2.9%	2.5%	2.8%
	Solano	North	21,470	27,815	35,417	44,275	11.9%	9.2%	10.0%
	Sonoma	North	26,762	34,494	43,614	54,514	14.5%	11.0%	12.3%
	Subregional Total		**72,618**	**88,760**	**108,673**	**132,573**	**30.3%**	**24.1%**	**26.9%**
	Monterey	South	20,224	28,922	50,837	74,896	16.3%	26.5%	27.1%
	San Benito	South	2,709	4,344	7,240	10,457	3.1%	3.5%	3.6%
	San Luis Obispo	South	14,989	20,920	32,512	45,581	11.1%	14.0%	14.7%
	Santa Cruz	South	11,539	14,713	21,142	21,145	6.0%	7.8%	0.0%
	Subregional Total		**49,461**	**68,899**	**111,731**	**152,079**	**36.5%**	**51.8%**	**45.4%**
	REGIONAL TOTAL		**344,764**	**398,027**	**480,638**	**569,534**	**100.0%**	**100.0%**	**100.0%**

Appendix 1. continued

Major Region	County	Sub-region	Urbanized Land Area (Hectares)				County Share of Regional Change		
			1998	2020F	2050F	2100F	1998–2020	2020–2050	2050–2100
San Joaquin Valley	Merced	North	12,358	18,528	29,353	41,382	7.6%	6.7%	6.8%
	San Joaquin	North	29,023	43,284	63,652	85,114	17.6%	12.7%	12.2%
	Stanislaus	North	20,430	25,142	38,362	54,256	5.8%	8.2%	9.0%
	Subregional Total		**61,811**	**86,954**	**131,367**	**180,752**	**31.1%**	**27.6%**	**28.0%**
	Fresno	South	37,765	48,893	81,243	119,323	13.7%	20.1%	21.6%
	Kern	South	40,840	65,117	111,187	159,400	30.0%	28.7%	27.3%
	Kings	South	11,501	15,094	20,830	27,189	4.4%	3.6%	3.6%
	Madera	South	9,025	15,348	24,970	35,827	7.8%	6.0%	6.2%
	Tulare	South	19,701	30,181	52,627	76,216	12.9%	14.0%	13.4%
	Subregional Total		**118,832**	**174,633**	**290,857**	**417,955**	**68.9%**	**72.4%**	**72.0%**
	REGIONAL TOTAL		**180,643**	**261,587**	**422,224**	**598,707**	**100.0%**	**100.0%**	**100.0%**
Sacramento	Sacramento	Central	61,009	71,950	85,317	100,003	35.5%	35.9%	34.0%
	El Dorado	Foot-hills	10,436	13,920	17,829	22,611	11.3%	10.5%	11.1%
	Nevada	Foot-hills	5,924	7,935	9,905	12,367	6.5%	5.3%	5.7%
	Placer	Foot-hills	15,284	23,776	33,099	44,089	27.5%	25.0%	25.5%
	Subregional Total		**31,644**	**45,631**	**60,833**	**79,067**	**45.3%**	**40.8%**	**42.2%**
	Sutter	North	4,311	6,385	9,202	12,333	6.7%	7.6%	7.3%
	Yuba	North	4,531	5,821	7,834	10,290	4.2%	5.4%	5.7%
	Subregional Total		**8,842**	**12,206**	**17,036**	**22,623**	**10.9%**	**13.0%**	**12.9%**
	Yolo	West	10,368	12,923	16,752	21,422	8.3%	10.3%	10.8%
	REGIONAL TOTAL		**111,863**	**142,710**	**179,938**	**223,115**	**100.0%**	**100.0%**	**100.0%**

References

Association of Bay Area Governments. 2000. *Projections 2000.* Oakland, Calif.

California Department of Conservation. 1998. California Farmland Mapping and Monitoring Program (CFMMP) data files.

Federal Emergency Management Agency (FEMA). 1996. Q3 flood maps (digital form). Washington, D.C.

Landis, John. 2001. Characterizing urban land capacity: Alternative approaches and methodologies. In *Land market monitoring for smart urban growth*, edited by G. J. Knaap. Cambridge, Mass: Lincoln Institute of Land Policy. 3–52.

Landis, John, Michael Reilly, Robert Twiss, Howard Foster, and Patricia Frontiera. 2001. Forecasting & mitigating future urban encroachment adjacent to California military installations: A spatial approach. Berkeley, Calif: Institute of Urban and Regional Development Working Paper #WP-2001-11.

Landis, John, Chris Cogan, Pablo Monzon, Michael Reilly. 1998. *Development and pilot application of the California urban and biodiversity analysis (CURBA) model.* Berkeley, Calif: Institute of Urban and Regional Development Monograph #MG-1998-01.

McFadden, Daniel. L. 1974. Conditional logit analysis of qualitative choice behavior. In *Frontiers in econometrics*, edited by P. Zarembka. New York: Academic Press. 105–142.

Sandoval, Juan Onesimo and John Landis. 2000. *Estimating the housing infill capacity of the Bay Area.* Berkeley, Calif.: Institute of Urban and Regional Development. Working Paper.

Endnotes

[1] The use of logit models to analyze discrete choices at a single point in time is firmly grounded in microeconomic theory (McFadden 1974). The use of logit models to analyze discrete land-use changes, particular changes identified from maps—while statistically feasible— introduces additional theoretical complications. In order for the estimated model parameters to be reliable—that is, to be free from bias—we must make two assumptions about the process of land-use change itself. The first is that all participants in the land development process must act independently of each other. This includes landowners, developers, builders, brokers, homebuyers, renters, and businesses. This assumption is intended to rule out the possibility of oligopolistic or strategic behavior. A second assumption concerns the lack of presence of any identifiable participants or agents. Discrete choice analysis has traditionally been used to model the behavior of identifiable agents such as voters, travelers, and consumers. In the case of land-use change, the agents of interest are land buyers and sellers. Models like the one identified above are known as reduced-form models because they include information on transac-

tion outcomes but not on the agents involved in the transaction. In simple economics terms, there are no utility-maximizing buyers or profit-maximizing sellers present in the model to start or complete a transaction. This is only a problem to the extent that the characteristics of specific buyers and sellers might affect their actions. To deal with this problem, we invoke the idea of competition. Specifically, we argue that if land market are competitive (e.g., there are no barriers to entry), then the characteristics and noneconomic motivations of particular agents should not affect transaction outcomes. Whether developers are well-capitalized or poorly-capitalized, whether they specialize in residential development or retail development, whether their experience is local or national—in a competitive market, these factors should be of less importance than the strength of the demand for urban development and the availability of appropriate sites.

[2] In situations where views are rewarded in the marketplace with price and rent premiums, the probability of development may actually rise with slope.

[3] This approach requires defining regions in terms of commute sheds.

[4] Most regional economic studies divide employment into (1) *basic jobs*, which generate income to a region or metropolitan area through the sales of goods and services to customers outside the region; and (2) *nonbasic*, or service jobs, which provide goods and services to the resident population and businesses with the region.

[5] By net job growth we mean the excess of job gains (through attraction and expansion) over losses (through contractions and firm death).

[6] DOF uses a modified cohort-component model, disaggregated by race and one-year age cohort to project population.

[7] The NWI does not include vernal pools.

Linking Land-use Change with Ecosystem Processes: A Hierarchical Patch Dynamic Model

Jianguo Wu, G. Darrel Jenerette, and John L. David

Introduction

Land-use and land-cover change is indicative of the power and will with which humans modify and conquer nature. Anthropogenic activities, such as agriculture and urbanization, have drastically transformed natural landscapes everywhere around the world, inevitably exerting profound effects on the structure and function of ecosystems. In the past 120 years, the world has lost 500 million hectares of forest to land conversion (Houghton et al. 1983, 1987; Ojima et al. 1994), while cultivation of grasslands in the Central Plains of the United States has resulted in losses of 800–2000 g C/m2 since settlement by Euro-American farmers (Burke and Schimel 1990; Burke et al. 1991). In particular, the conversion of natural and agricultural areas to highly artificially modified urban land uses has been taking place at an astonishing rate. According to the United Nations, the world urban population was only a few percent of the global population in the 1800s, but increased to nearly 30 percent in 1950 and reached 50percent in 2000. It has been projected that 60percent of the world population will live in urban areas by 2025. By contrast, urban land in the United States increased by 22 million acres between 1960 and 1980 (Frey 1984), that is, by 1.1 million acres per year. Seventy-four percent of the U.S. population (203 million people) resided in urban areas in 1989, and that number is projected to rise to more than 80 percent in 2025 (Alig and Healy 1987; McDonnell et al. 1997).

Although urban areas represent arguably the most important habitats for humans, they are among the least understood ecosystems of all, and urban ecology has not been considered part of the mainstream ecology worldwide (Collins et al. 2000; Wu 2001). It is true that ecological studies in urban areas have a long history dating back to the early 1900s or earlier (see Harshberger 1923; Breuste et al. 1998). In parallel, much research has been done in spatial pattern and urban dynamics by geographers and social scientists with little or only superficial consideration of ecology in and around cities (e.g., Forrester 1969; Berry and Kasarda 1977; Batty and Longley 1994). However, understanding how urban ecosystems work does not come simply from a large number of botanical, zoological, sociological, or geographic investigations within cities. The urban whole is larger than the sum of its biological and abiotic parts. The ecology of urban systems as integrated wholes needs new and integrative perspectives (Pickett et al. 1997; Grimm et al. 2000; Zipperer et al. 2000).

In this paper, we describe the theoretical basis and the general structure of a hierarchical patch dynamics model for the Phoenix metropolitan region, HPDM-PHX, which integrates land use change with ecosystem processes. The main goal of our modeling project is to develop an understanding of how urbanization affects ecosystem productivity and biogeochemical cycles at local and regional scales. In particular, the model is used to address the following questions: How has the landscape pattern of the Phoenix metropolitan area been transformed by agriculture and urbanization since the early 1900s? How have these land use and land cover changes affected ecosystem production and nutrient cycling (e.g., C and N)? How do primary production and carbon and nitrogen dynamics differ among natural, agricultural, and urban ecosystems along a landscape gradient of urbanization? How do variations in climatic conditions (precipitation and temperature) affect the primary production and C and N dynamics for different land cover types? As the project is still unfolding, this paper focuses primarily on the theoretical basis and general structure of the hierarchical patch dynamics model of the Phoenix metropolitan area.

Land-use Change and Ecosystem Processes

Land-use and land-cover changes may significantly affect the composition of plant communities by fragmenting the landscape, removing and introducing species, and altering water, carbon, and nutrient pathways (Ojima et al. 1994; Vitousek et al. 1997; Vitousek and Mooney 1997; Shugart 1998). These changes may result in modifications of land surface properties, such as surface albedo and latent and sensible heat fluxes (Pielke and Avissar 1990), and also modify the quality and quantity of litter and allocations of carbon and nutrients, further enhancing greenhouse gas feedback to climate systems (Hall et al. 1988; Ojima et al. 1994). Therefore, land use and land cover changes and their influences on ecosystem processes must be incorporated to address large-scale ecological and environmental issues such as urbanization, global climate change, desertification, and resource management. Recent studies in landscape ecology have indicated that understanding the interactions between landscape pattern and ecological processes at broad spatial scales is crucial for properly managing natural and human-dominated ecosystems (Moss 1988; Risser 1990; Ludwig et al. 1997; Dale et al. 2000).

To investigate the ecological consequences of land use and land cover changes, a spatially explicit, landscape ecological approach is essential. On the one hand, land use and land cover change is inherently a spatial process, and simulating land use and land cover change must consider neighborhood effects that represent local-scale interactions as well as top-down constraints imposed from broader scales. On the other hand, studies of ecosystem processes need to incorporate landscape patterns that vary in both space and time. Forman (1990) stated that "for any landscape, or major portion of a landscape, there exists an optimal spatial configuration of ecosystems and land uses to maximize ecological integrity, achievement of human aspirations, or sustainability of an environment." We speculate that there may be multiple spatial configurations that are equally optimal in complex spatial systems such as urban landscapes, although it would be extremely difficult, if ever possible, to test such hypotheses through field experiments. Nevertheless, empirical studies have demonstrated that the configuration of landscape elements (e.g., natural vegetation remnant patches, parks, golf courses, agricultural fields, and urban blocks) often influences various ecosystem processes, such as net primary productivity, watershed discharge characteristics, and nutrient cycling (Lowrance et al. 1985; McDonnel and Pickett 1990; Risser 1990; Knapp et al. 1993; McDonnel et al. 1997). Recent ecological studies also have suggested that measures of landscape pattern (indices or metrics) may reveal ecological processes operating at different scales (e.g., Krummel et al. 1987; Hoover and Parker 1991; Graham et al. 1991; Hunsaker et al. 1994; Wu et al. 2000). While caution must be carefully taken, landscape structural measures may be used as indicators for monitoring ecosystem changes at regional scales (O'Neill et al. 1994, 1997; Jones et al. 1996). Thus, methods of spatial pattern analysis not only are important for quantifying landscape structure and its change, but for relating landscape pattern to ecological processes as well (Wu and Qi 2000).

Urbanization and Landscape Pattern Change in Phoenix

The Phoenix metropolitan region, Arizona, the United States, is located in the northern part of the Sonoran Desert. Phoenix is the home of the Central Arizona–Phoenix Long-Term Ecological Research (CAPLTER) Project, supported by the U.S. National Science Foundation. The major goal of CAPLTER is to understand the pattern and process of urbanization and their ecological consequences. While the climate of this area is one of the hottest and driest regions in North America, the biodiversity of the Sonoran Desert is among the richest of all deserts in the world. Summer temperatures average 30.8oC while winters are warm with average temperatures of 11.3oC. Average annual precipitation in the Phoenix area is 180 mm, with approximately 50 percent in the form of summer thunderstorms and the remainder associated with winter frontal systems originating in the Pacific Ocean. Native vegetation is characterized by desert scrub communities that are dominated by creosote bush (*Larrea tridentata*), mesquite (*Prosopis glandulosa*), and several other shrub species. The giant cactus, saguaro (*Carnegiea gigantea*), standing tall with multiple arms reaching out, is found throughout the area either as the prominent landscaping plant in human-constructed environments or as the monarch of a variety of cactus plants in the desert. With its magnificent charisma and sacred status, saguaro is undoubtedly the most recognizable symbol of the Sonoran Desert landscape (see photo, Figure 1).

Figure 1. Historical land-use change in the Phoenix metropolitan area between 1912 and 1995 (modified from Knowles-Yanez et al. 1999).

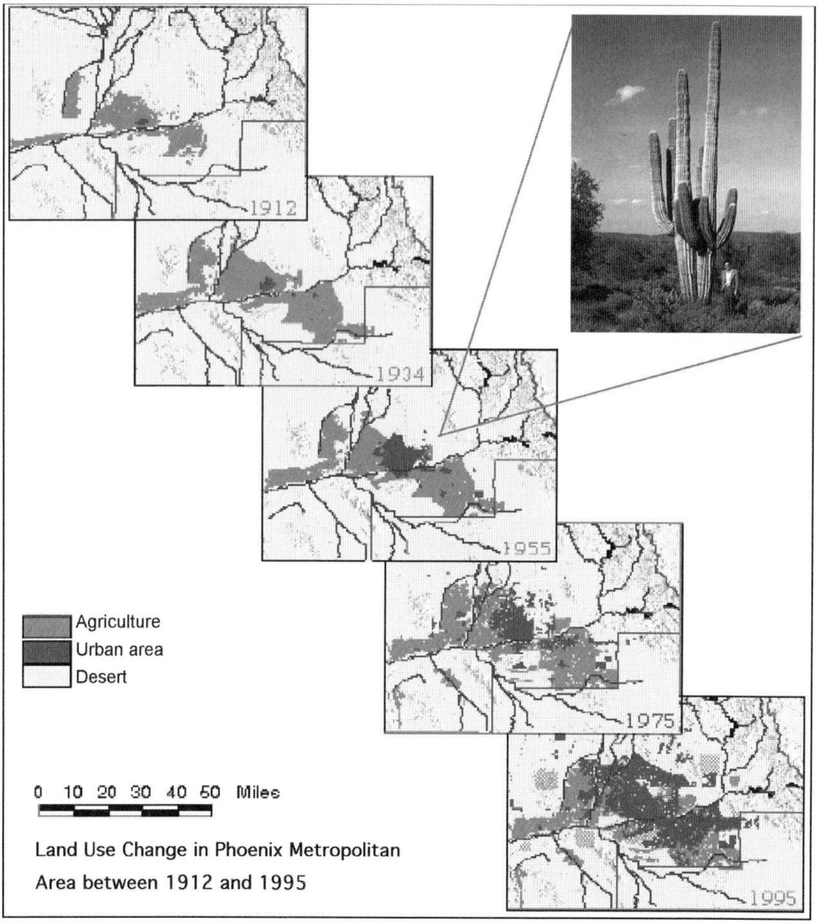

In the southwest United States and the Phoenix metropolitan area in particular, urbanization has rapidly transformed the desert landscape (Figure 1). According to the U.S. Census Bureau, Arizona had a net percent population increase of 30.4 between 1990 and 1999, second only to Nevada, whose population increased by 50.6 percent during the same period. In recent decades, Phoenix has become one of the largest and fastest growing cities in the United States (Figure 2, Table 1), with more than a half of the population of the entire state of Arizona concentrated in the Phoenix metropolitan area. The population of Maricopa County, where the Phoenix metropolitan area is located, was only 5,689 in 1880. It increased to 1.51 million in 1980 and reached 2.72 million in 1997. By contrast, the population of the state of Arizona was 4.55 million in 1997 and 4.67 million in 1998. The rapid

Figure 2. Population growth in Maricopa County of the Phoenix metropolitan area between 1880 and 1997. The population growth is highly correlated with the expansion of urbanized area.

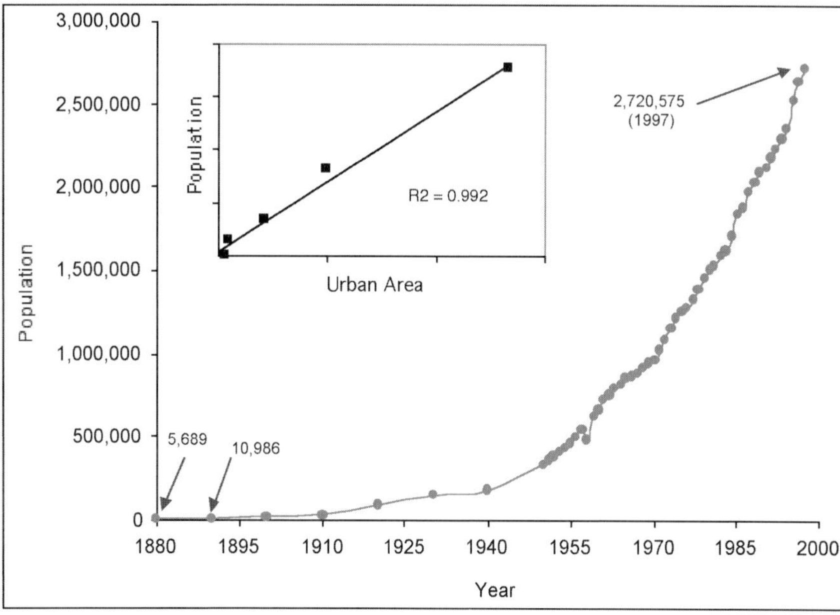

Table 1. The top 10 fastest growing metropolitan areas in the United States between 1990 and 1998 (data source: U.S. Census Bureau http://www.census.gov/).

	Rank by Population Size		Rank by Percent Change
	7/1/98	4/1/90	4/1/90 to 7/1/98
Las Vegas, Nev.	33	51	1
Laredo, Tex.	167	201	2
McAllen-Edinburg-Mission, Tex.	78	95	3
Boise City, Idaho	100	117	4
Naples, Fla.	161	177	5
Phoenix-Mesa, Ariz.	14	19	6
Austin-San Marcos, Tex.	41	52	7
Fayetteville-Springdale-Rogers, Ark.	134	148	8
Wilmington, N.Carolina	151	166	9
Provo-Orem, Utah	117	129	10

population growth in the Phoenix metropolitan area has led to an equally fast expansion of urbanized area. The tight linear correlation between population and urban area (Figure 2; see Jenerette and Wu 2001) suggests that rapid urban sprawl in the Phoenix area will persist at least in the near future as a result of the continued explosive population growth due to the influx of immigrants.

Urbanization has resulted in dramatic structural changes of the Sonoran Desert landscape. For example, as urbanization progressed large, contiguous desert patches were broken up (Figure 3A, B), with an increasing number of patch types (land use types) occurring in the landscape. The density of patches of various types and thus the density of edges increased exponentially (Figure 3C, D). The overall patch diversity increased steadily due mainly to the increasingly even proportions of dominant land use types (Figure 3E), whereas the geometric shapes of patches in the landscape as a whole became more and more irregular (Figure 3F). In short, urbanization has made the Phoenix landscape structurally more fragmented and complex. These land use and land cover changes due to urbanization inevitably result in significant alterations of the biological composition and spatial configuration of local ecosystems, which in turn have important effects on water, carbon, and nutrient cycles, and the climatic systems at the landscape and regional scales.

Model Structure of HPDM-PHX

Theoretical basis

Our modeling approach is based on the concepts and principles of the hierarchical patch dynamics paradigm, which integrates hierarchy theory with the theory of patch dynamics (Wu and Levin 1994; Wu and Loucks 1995; Reynolds and Wu 1999; Wu 1999). The complexity of ecological systems stems from the multiplicity of spatial patterns and ecological processes, nonlinear interactions among components, and spatial heterogeneity (O'Neill et al. 1986; Wu 1999). Simon (1962) noted that complexity frequently takes the form of hierarchy, whereby a complex system consists of interrelated subsystems that are in turn composed of their own subsystems, and so on, until the level of elementary or "primitive" components is reached. A major utility of hierarchy theory is to simplify complexity (deriving order out of seeming disorder) and thus facilitating prediction and understanding. In the case of building complex yet stable software systems, computer scientists have developed the object-oriented design, analysis, and programming paradigm following a hierarchical approach (Booch 1994). On the other hand, effective human problem-solving procedures also are hierarchical (Newell and Simon 1972). It has been argued that if a complex system is not hierarchical, we may never be able to adequately describe it; if we could, it would be hardly comprehensible (Simon 1973).

Figure 3. Structural changes of the Phoenix metropolitan landscape between 1912 and 1995. Rapid urbanization has resulted in an accelerating increase in landscape fragmentation and structural complexity.

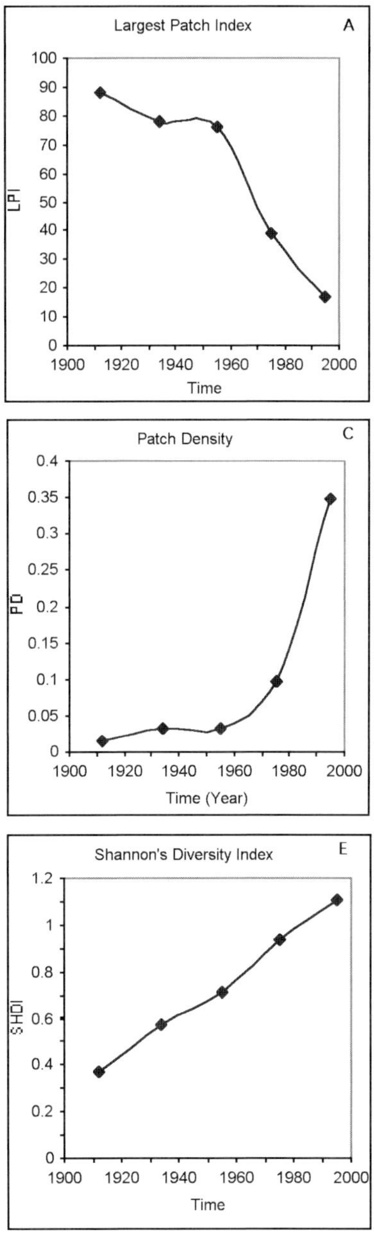

Table 2. Main components of hierarchical patch dynamics paradigm (see Wu and Loucks 1995 and Wu 1999 for details).

- Ecological systems are spatially nested patch hierarchies, in which larger patches are made of smaller patches.
- Dynamics of a given ecological system can be viewed as the composite dynamics of patches at adjacent hierarchical levels.
- Pattern and process interact with each other and their relationship changes with scale.
- Nonequilibrium and stochastic processes are common in ecological systems and can be forces that lead to order or organization at broader scales.
- Persistent ecological systems usually exhibit metastability, which is often achieved through spatial incorporation.

Spatial patchiness is ubiquitous in nature, and patch dynamics are common and essential in many ecological systems. A fundamental flaw in the classic equilibrium paradigm in ecology has been its inability to recognize the importance of heterogeneity and scale linkages of patterns and processes. The recent emphasis transition in ecology, from equilibrium, homogeneity, determinism, and local or single-level phenomena to non-equilibrium, heterogeneity, stochasticity, and hierarchical properties, clearly indicates a paradigm shift in ecology. The hierarchical patch dynamics paradigm explicitly emphasizes the dynamic relationship among pattern, process, and scale in a landscape context (see Table 1). While hierarchy theory provides useful guidelines for "decomposing" complex systems by giving them a "vertical" structure, patch dynamics emphasizes the dynamics of spatial heterogeneity and horizontal interactions between patches in a landscape (Wu and Levin 1994; Wu and Loucks 1995; Wu 1999).

To "decompose" the complexity of the Phoenix metropolitan landscape in terms of its structure and functionality, we adopt the concept of ecosystem functional types (EFTs) (see figure 4), which provides a way of grouping a large number of local ecosystems into a smaller number of categories that each have similar functional properties in terms of biogeochemical cycling (Reynolds et al. 1997; Reynolds and Wu 1999). Thus, EFTs in modeling the functioning of landscapes are similar to trophic levels or guilds in modeling foodweb dynamics. We distinguish three EFTs hierarchically at three distinctive spatial scales: the local ecosystem, landscape, and region (Figure 3). Because the EFT concept emphasizes ecosystem attributes and processes (e.g., primary productivity, biogeochemical cycling, gas fluxes, hydrology), it provides concrete meanings to patches and thus reinforces the less tangible process aspect of the hierarchical patch dynamics paradigm. While these EFTs possess spatial heterogeneity at different spatial and temporal scales, we hypothesize that changes in ecosystem structure and function are much smaller within each patch type than between patch types, and that these changes show detectable scale discontinuities in space (Wu et al. 2000). This hierarchical EFT approach facilitates our understanding of the diversity and distribution of local ecosystems as well as their changes due to urbanization in the region. In addition, it allows us to model similar local ecosystems with the same model structure and similar sets of parameters.

Figure 4. Hierarchical ecosystem functional types for the Phoenix metropolitan area. The EFT hierarchy consists of the local ecosystem, landscape, and region levels, each of which is characterized by a set of distinct features in structure and function.

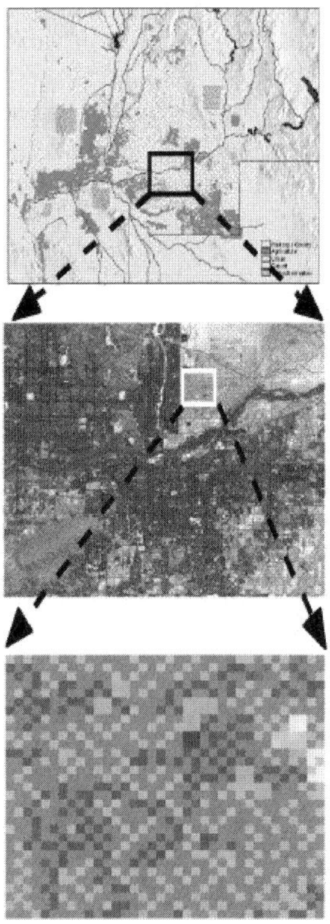

EFT/Scale
Major Characteristics

Regional EFT (the regional landscape)
- Composed of different types of local landcapes
- Heterogeneous in ecosystem structure and function
- Characterized by the dominant biome and land use type at the regional geographical scale (e.g., an urbanized desert region vs. an agricultural grassland region)

Landscape EFTs (local landscapes)
- Composed of different land use and land cover types
- Heterogeneous in ecosystem structure and function
- Characterized by dominant land use types, such as urban landscapes, rural landscapes, agricultural landscapes, and natural desert landscapes

Local EFTs
(local ecosystems or land cover types)
- Relatively homogeneous vegetation-soil complexes
- Readily detectable from air photos and remote sensing data (e.g., Landsat images)
- Largely corresponds to the categories of the Anderson et al. (1976) level II classification

We define a local EFT (or local ecosystem) as a land use/land cover type with a relatively homogeneous vegetation-soil complex (e.g., an agricultural field, a residential area, a park, a remnant desert fragment). Such local EFTs are readily detectable from air photos and remote sensing data (e.g., Landsat images), and they largely correspond to the categories of the Anderson et al. (1976) level II classification. The landscape EFTs (or local landscapes) are spatial mosaics of a number of local EFTs of different types. They are heterogeneous in ecosystem structure and function, and each dominated by one or a few land use/land cover types. For example, urban landscapes are filled with human constructions, agricultural landscapes are replete with cultivated fields, and natural desert landscapes are dominated by native vegetation. Conceivably, the structure and function of a landscape EFT is a function of both the landscape composition (the variety of patch types and their relative abundance) and the configuration of patches (e.g., patch shape and spatial arrangement). These characteristics are important in determining the behavior of a particular mosaic (e.g., the effect of vegetation on hydrologic flow [Pickup 1985; Turner and Garnder 1991; Wondzell et al. 1996] and exchanges of water, organic matter, propagules, nutrients, sediments, etc. [Sklar and Constanza 1991]. A regional EFT is a mixture of the landscape EFTs or local landscapes, and characterized by the dominant biome and land use type at the regional geographical scale. In our case, the regional EFT is the Phoenix metropolitan region of the Sonoran Desert.

The local, landscape, and regional EFTs provide a hierarchical structure to the system under study and an integrative framework for coupling landscape pattern with ecosystem processes (e.g., biogeochemical cycles). Patch dynamics occur simultaneously over a range of scales at differential rates, and our hierarchical patch dynamics model aims to scale up ecological processes from the local ecosystem to the landscape and then the regional level in the spatially nested hierarchy. Thus, HPDM-PHX has three distinctive hierarchical levels built in its structure: the local ecosystem, the landscape, and the region.

Local ecosystem model

At the local ecosystem level, we use modified versions of two ecosystem process models: CENTURY, a general model of terrestrial biogeochemistry originally developed for the Great Plains grassland ecosystem by Parton et al. (1987, 1988) and PALS, a patch-level arid ecosystem simulator developed by J. F. Reynolds and associates for the Jornada Basin, New Mexico (Reynolds et al. 1993, 1997).

CENTURY simulates the long-term dynamics of carbon, nitrogen, phosphorus, sulfur, and plant production and has been tested for a number of grassland ecosystems worldwide (Parton et al. 1993). The main input data for CENTURY include: monthly average maximum and minimum air temperature, monthly precipitation, lignin content of plant materials, plant C and N, soil texture, atmospheric and soil N inputs, and initial soil C and N levels. Model output includes information on carbon and nitrogen fluxes, net primary production, and soil organic matter. While it contains several submodels, the main governing equations in the CENTURY model are as follows (Parton et al. 1993):

$$\frac{dC_i}{dt} = K_i L_C A C_i \qquad i = 1, 2$$

$$\frac{dC_i}{dt} = K_i A T_m C_i \qquad i = 3$$

$$\frac{dC_i}{dt} = K_i A C_i \qquad i = 4, 5, 6, 7, 8$$

$$P_p = P_{mx} T_p M_p S_p$$

where C_i is the carbon in the state variable; i =1, 2, 3, 4, 5, 6, 7, 8 denote surface and soil structural material, active soil organic matter, surface microbes, surface and soil metabolic material, slow and passive soil organic matter fractions; K_i is the maximum decomposition rate (year^{-1}) for the ith state variables; A is the combined abiotic impact of soil moisture and soil temperature on decomposition (product of the soil moisture and temperature terms); P_p is the aboveground potential plant production rate (g m^{-2} month^{-1}); P_{mx} is the maximum potential aboveground plant production rate; T_p is the effect of soil temperature on growth; M_p is the effect of moisture on production; and S_p is the effect of plant shading on plant growth.

PALS simulates carbon, water, nitrogen, and phosphorus cycles, and takes into account variations in patch type, plant characteristics, soil resources, and climatic factors (Figure 5). The abiotic components of PALS include micrometeorological conditions (e.g., temperature and moisture within and above the canopy) and soil properties (e.g., water flux, nutrients, temperature). The model PALS is well suited to explore questions related to nutrient cycling and has been parameterized for the Jornada LTER site, California chaparral, and a grassland in Kansas (Reynolds et al. 1997, 1999). Main advantages of using PALS for our project in Phoenix include: (1) it includes the major ecosystem processes in desert systems, (2) the model has been tested on several sites, and (3) the similarity in dominant plant species in the Jornada Basin and the Phoenix area means that model parameterization can be greatly facilitated. We use these two ecosystem models in parallel for the following reasons. CENTURY and PALS represent different levels of mechanistic details in simulating ecosystem processes, and thus comparing them can help us understand what details can be ignored in the process of scaling up from the local ecosystem to the region. Model comparison provides a means for increasing our confidence in estimating ecological variables especially when data are rarely available (Schimel et al. 1997). Moreover, ecosystem models that are tailored for different land cover types encountered in the Phoenix metropolitan area need to be developed based on CENTURY and PALS. We are currently in the process of collecting and compiling input data for running CENTURY and PALS for several major land cover types in the Phoenix metropolitan area. These models will be validated and compared at the local ecosystem level and among different local EFTs before being spatially incorporated into the landscape and regional models.

Land-use and land-cover change model

The applications of spatial Markovian and cellular automata approaches in modeling vegetation dynamics, land use and land cover change, and urban growth have mushroomed in the past two decades (e.g., Couclelis 1985; Turner 1987; Batty and Xie 1994; Green 1994; Clarke et al. 1997; Wu 1998). The combination of the two approaches is often desirable when stochastic factors are important in determining local transition rules (e.g., Li and Reynolds 1997; Balzter et al. 1998). However, these models usually are not integrated or coupled with ecosystem models. While our land use and land cover (sub)model in HPDM-PHX shares some of

Figure 5. Illustration of the plant functional type based ecosystem model, PALS (Reynolds et al. 1993, 1997), and its role in the scaling up of ecological processes.

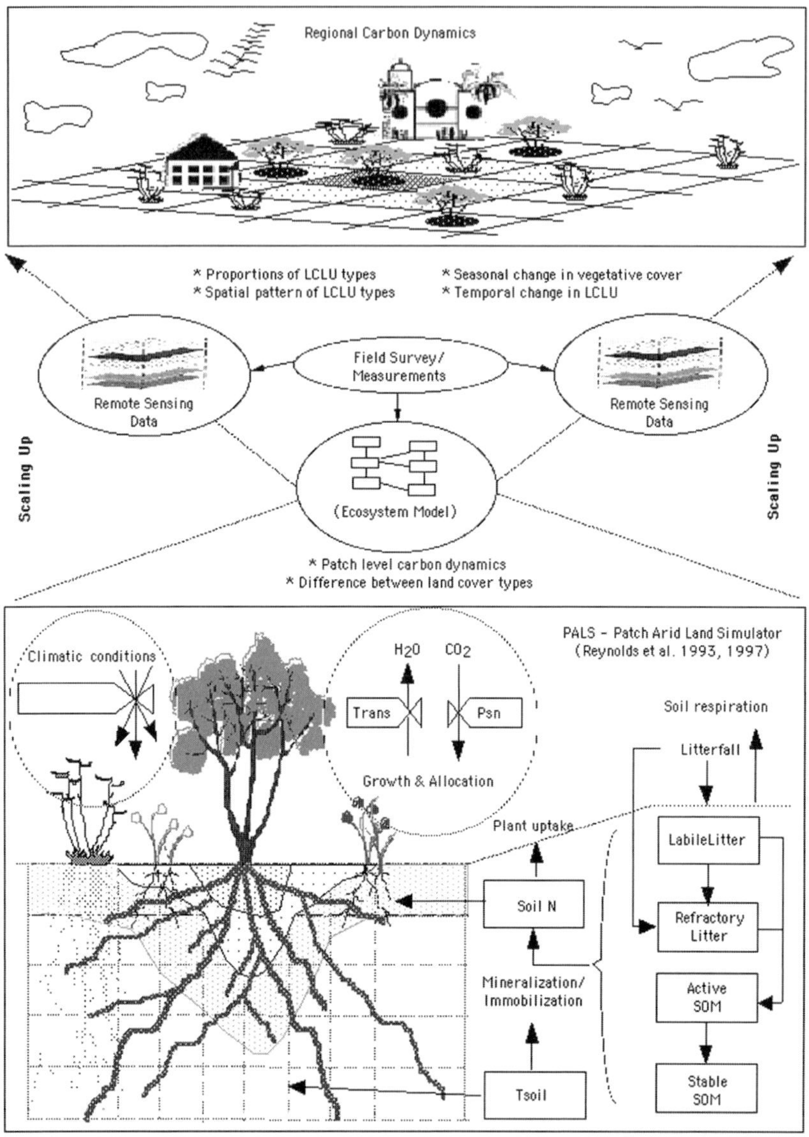

the similarities of the Markov–cellular automata approach, it is fully integrated with the ecosystem model. The landscape and regional models conceptually resemble each other, but differ significantly in spatial extent—the landscape model is nested in the regional model. The regional model is the integration, rather than a simple summation, of various component landscapes when horizontal interactions between them are strong and nonlinear.

We have developed two land use change models based on the historical land use change data for the Phoenix metropolitan area. The first was a Markov–cellular automata model (Jenerette and Wu 2001), in which parameters and neighborhood rules were obtained both empirically and with a genetic algorithm (GA). The model simulated the change in land use pattern better with the optimized parameter set using GA than with the empirically derived parameter set. While a high degree of accuracy of statistical properties of the simulated results was readily achieved, the spatial structure of land use patterns was only satisfactory at coarse scales. To improve the spatial accuracy, we have developed a hierarchical land use and land cover change model that takes into account both the local neighborhood effects and influences and constraints at broader spatial scales (e.g., ownership and administrative boundaries). The incorporation of information on ownership substantially improved the overall accuracy of the simulated land use pattern (Figure 6; David and Wu 2000). The current version of the model simulates only three land use types. However, to link it with ecosystem process models, we still need to modify the land use change model to represent several important land cover types in the metropolitan area.

The land use and land cover model and ecosystem models are integrated through a framework illustrated in Figures 7 and 8. The land use change model is driven by local rules and top-down constraints which are in turn influenced by socioeconomic processes in the region. Changes in landscape pattern then result in changes in ecosystem processes at both local and regional scales. Although the effects of land use change on ecological processes are often more obvious and dominant than the feedback of changed ecological conditions to land use decisions, the latter does exist and will become more important as urbanization continues to progress. Model validation and applications will involve several steps: 1) to assess the reasonableness of the model structure and the interpretability of functional relationships within HPDM-PHX; 2) to simulate ecosystem processes across a gradient of land cover types; 3) to evaluate the correspondence between model behavior and the expected patterns of model behavior at local ecosystem, landscape, and regional scales; and 4) to conduct a series of sensitivity and uncertainty analysis with HPDM-PHX.

Figure 6. Land-use change in the Phoenix metropolitan area. Top: 1995 land-use classification map; middle: simulated land-use pattern for 1995; and bottom: projected land-use pattern for 2030.

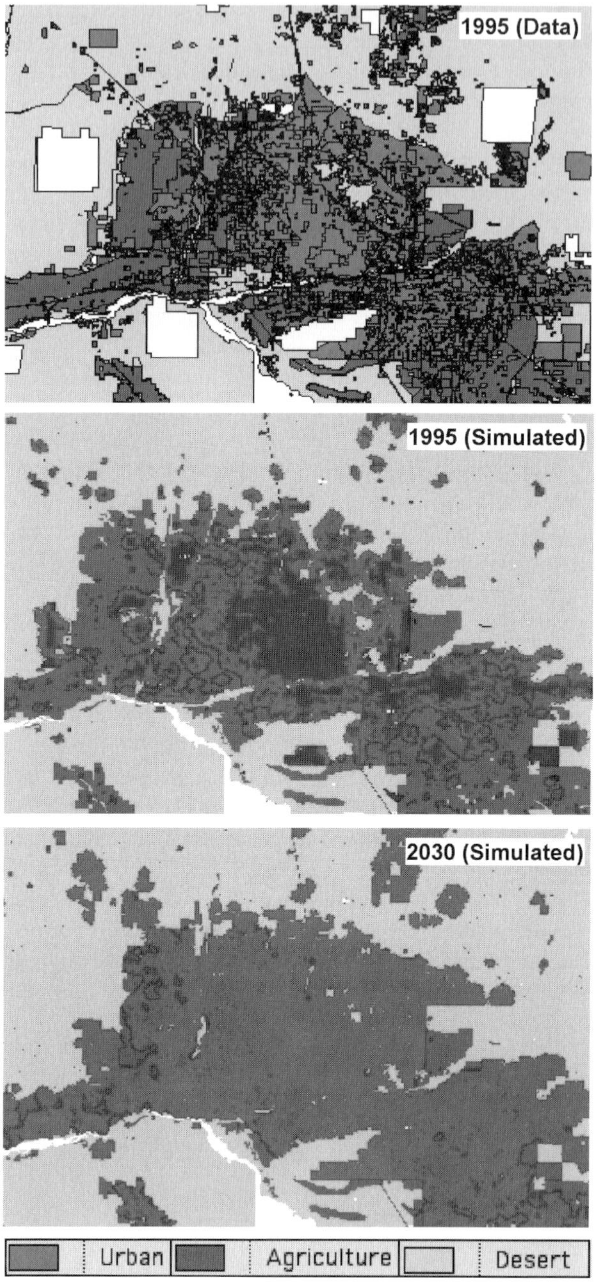

Figure 7. Illustration of the coupling between the ecosystem model and land-use change model within a GIS framework.

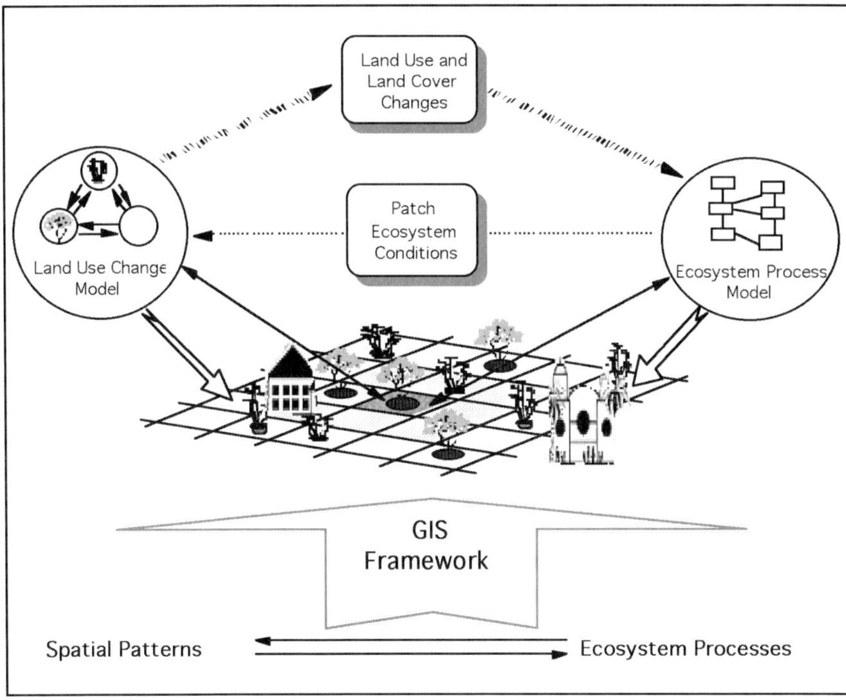

Discussion and Conclusions

Land-use change is perhaps the most conspicuous and pervasive human alteration to the surface of the earth. Although the Great Wall of China may look more spectacular from the space, land use change in the forms of urbanization and agriculture has far more profound and widespread ecological consequences. Indeed, habitat destruction is generally identified as the major cause for the loss of biodiversity and habitat destruction occurs mostly in the form of land conversion. There is no other form of land transformation that alters natural environments more radically than urbanization, which is an important global change problem that has received much less attention from either scientists or decision makers as compared to issues of global climate change. It only becomes evident in recent years that land use change is important to regional and global climate change (Houghton 1994; Houghton et al. 1999).

Undoubtedly, rapid global urbanization continues to have significant impacts on the environment as well as on economic, social, and political processes at local,

Figure 8. Schematic representation of the model structure of HPDM-PHX.

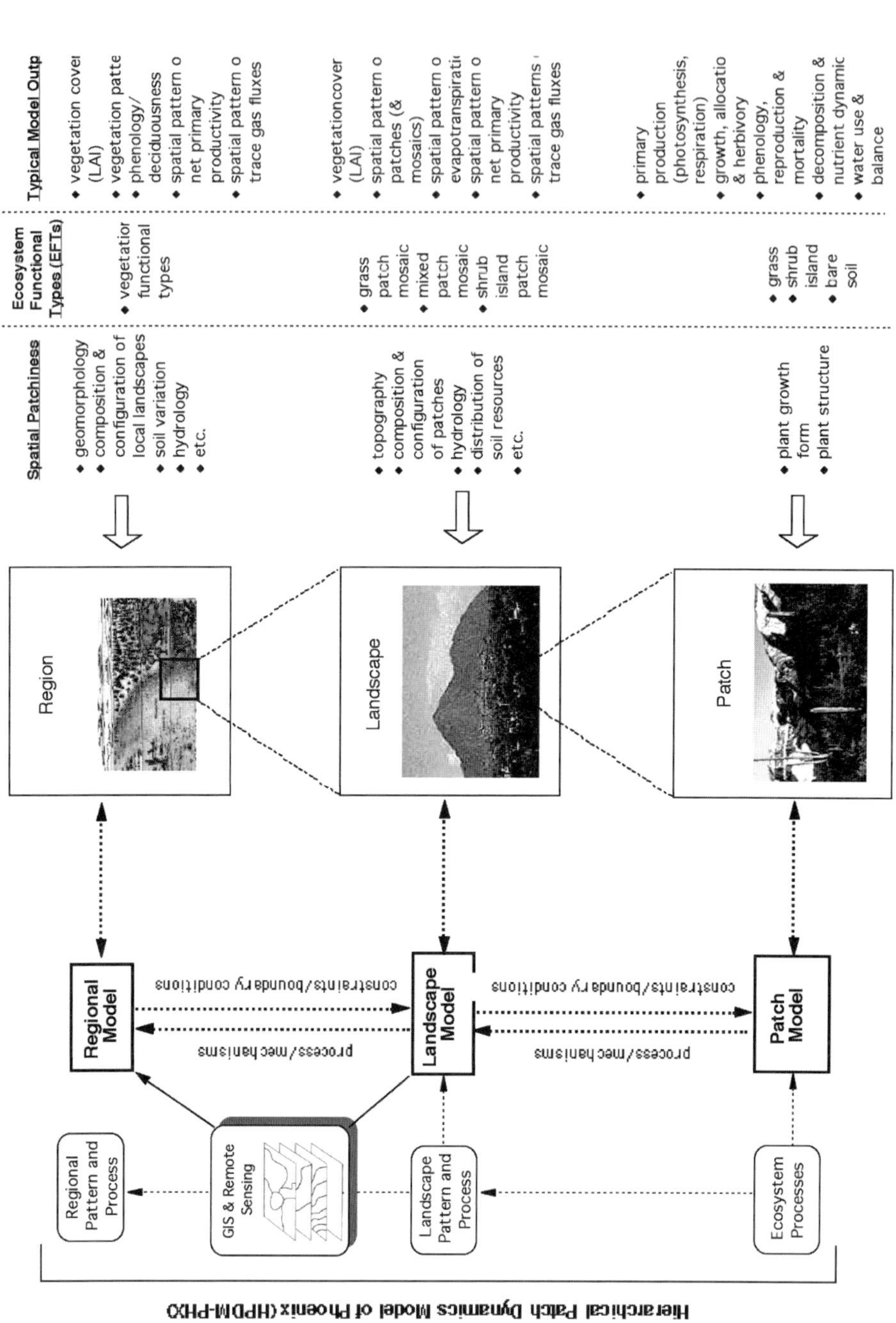

regional, and global scales. While the urban environment represents one of the most critical habitats for the survival and civilization of modern humans, they are among the least studied and most poorly understood. One may argue that urban ecology as "ecology in cities" or "human ecology" or "social ecology" in urban areas is as old as ecology itself. But much of the previous research in urban ecology has been more partial than comprehensive, more descriptive than explanatory, and more disciplinarily biased than interdisciplinarily integrated. More comprehensive, integrative perspectives that explicitly consider both ecological and socioeconomic components and their interactions in urban systems are needed. Urban environments exhibit arguably the most conspicuous and complex spatial heterogeneity, which often appears to be hierarchical, and a landscape ecology perspective is thus essential for studying the ecology of cities (Zipperer et al. 2000). We need to understand urban systems as integrative landscapes, i.e., dynamic patch mosaics that are created, modified, maintained, and destroyed by ecological and socioeconomic processes. Undoubtedly, interactions between pattern and process at different scales in urban landscapes may frequently lead to emergent properties that can not be understood by focusing only on individual patches.

Here, we present a hierarchical patch dynamics model, HPDM-PHX, that deals explicitly with spatial heterogeneity, functional complexity, and scale multiplicity in the Phoenix urban landscape. The model is based on the hierarchical patch dynamics paradigm (Wu and Loucks 1995) and the hierarchical scaling ladder approach (Wu 1999). A salient feature of the spatially explicit hierarchical model is that it integrates land use and land cover change with ecosystem processes explicitly at different spatial scales. Although developed for a particular urban landscape, the modeling approach should be applicable to other landscapes of different types. With this model, we hope to effectively address the question: How does urbanization affect the landscape structure and ecosystem processes in the Phoenix metropolitan area?

Solutions to ecological and environmental problems entail understanding and prediction of natural and anthropogenic patterns and processes on broad spatial and temporal scales. However, most ecological studies have been conducted on fine scales and as a consequence our knowledge of our environment also is polarized toward local scales. Thus, a grand challenge for regional scale analysis and assessment is to unravel how spatial heterogeneity at coarse scales affects ecological processes and to develop scaling strategies and rules for extrapolating information from the local ecosystem to the landscape and to the region. We believe that the hierarchical patch dynamics modeling and scaling approach can facilitate the integration between disciplines and across scales in the study of regional patterns and processes.

Acknowledgements

The research was in part supported by NSF grant DEB 97-14833 (CAP-LTER) and U.S. EPA grant R827676-01-0 (to JW). Although the research described in this article has been funded in part by the above mentioned agencies, it has not been subjected to the Agencies' required peer and policy review and therefore does not necessarily reflect the views of the agencies and no official endorsement should be inferred.

References

Alig, R. J., and R. G. Healy. 1987. Urban and built-up land area changes in the United States: An empirical investigation of determinants. *Land Economics* 63:215–226.

Anderson, J. R., E. E. Hardy, J. T. Roach, and R. E. Witmer. 1976. A land use and land cover classification system for use with remote sensor data. *Geological Survey Professional Paper 964*. Washington, D.C.: United States Government Printing Office.

Balzter, H., P. W. Braun, and W. Köhler. 1998. Cellular automata models for vegetation dynamics. *Ecological Modelling* 107:113–125.

Batty, M., and P. Longley. 1994. *Fractal cities: A geometry of form and function.* San Diego: Academic Press.

Batty, M., and Y. Xie. 1994. From cells to cities. *Environment and Planning B: Planning and Design* 21:s31–s48.

Berry, B. J. L., and J. D. Kasarda. 1977. *Contemporary urban ecology.* New York: Macmillan.

Booch, G. 1994. *Object-oriented analysis and design with applications.* 2nd ed. Reading: Addison-Wesley.

Breuste, J., H. Feldmann, and O. Uhlmann, eds. 1998. *Urban ecology.* Berlin: Springer.

Burke, I., and D. Schimel. 1990. Regional modeling of grassland biogeochemistry using GIS. *Landscape Ecology* 4:45–54.

Burke, I. C., T. G. F. Kittel, W. K. Lauenroth, P. Snook, C. M. Yonker, and W. J. Parton. 1991. Regional analysis of the central Great Plains. *BioScience* 41:685–692.

Clarke, K. C., S. Hoppen, and L. Gaydos. 1997. A self-modifying cellular automaton model of historical urbanization in the San Francisco Bay area. *Environment and Planning B: Planning and Design* 24:247–261.

Collins, J. P., A. Kinzig, N. B. Grimm, W. F. Fagan, D. Hope, J. Wu, and E. T. Borer. 2000. A new urban ecology. *American Scientist* 88:416–425.

Couclelis, H. 1985. Cellular worlds: A framework for modelling micro-macro dynamics. *Environment and Planning A* 17:585–596.

Dale, V. H., S. Brown, R. A. Haeuber, N. T. Hobbs, N. Huntly, R. J. Naiman, W. E. Riebsame, M. G. Turner, and T. J. Valone. 2000. Ecological principles and guidelines for managing the use of land. *Ecological Applications* 10:639–670.

David, J., and J. Wu. 2000. A hierarchical patch dynamics modeling platform. In *Proceedings of the International Conference on Modeling Complex Systems*, Montreal, edited by D. Marceau. Montreal: Univesity of Montreal.

Forman, R. T. T. 1990. Ecologically sustainable landscapes: The role of spatial configuration. In *Changing landscapes: An ecological perspective*, edited by I. S. Zonneveld and R. T. T. Forman. New York: Springer-Verlag.

Forrester, J. W. 1969. *Urban dynamics*. Cambridge: The M.I.T. Press.

Frey, H. T. 1984. *Expansion of urban area in the United States: 1960–1980*. U.S.D.A. Economic Research Service Staff Report No. AGES830615. Washington, D.C.

Graham, R. L., C. T. Hunsaker, R. V. O'Neill, and B. L. Jackson. 1991. Ecological risk assessment at the regional scale. *Ecological Applications* 1:196–206.

Green, D. G. 1994. Connectivity and complexity in landscapes and ecosystems. *Pacific Conservation Biology* 1: 194–200.

Grimm, N. B., J. M. Grove, C. L. Redman, and S. T. A. Pickett. 2000. Integrated approaches to long-term studies of urban ecological systems. *BioScience* 50:571–584.

Hall, F. G., D. E. Strebel, and P. J. Sellers. 1988. Linking knowledge among spatial and temporal scales: Vegetation, atmosphere, climate and remote sensing. *Landscape Ecology* 2:3–22.

Harshberger, J. W. 1923. Hemerecology: The ecology of cultivated fields, parks, and gardens. *Ecology* 4:297–306.

Hoover, S. R., and A. J. Parker. 1991. Spatial components of biotic diversity in landscapes of Georgia, USA. *Landscape Ecology* 5:125–136.

Houghton, R. A. 1994. The worldwide extent of land-use change. *BioScience* 44:305–312.

Houghton, R. A., J. L. Hackler, and K. T. Lawrence. 1999. The U.S. carbon budget: Contributions from land-use change. *Science* 285:574–578.

Houghton, R. A., J. E. Hobbi, J. M. Melillo, B. Moore, B. J. Peterson, G. R. Shaver, and G. M. Woodwell. 1983. Changes in the carbon content of terrestrial biota and soils between 1860 and 1980: A net release of CO2 to the atmosphere. *Ecological Monographs* 53:235–262.

Houghton, R. A., R. D. Boone, J. R. Fruei, J. E. Hobbie, J. M. Melillo, C. A. Palm, B. J. Peterson, G. R. Shaver, G. M. Woodwell, B. Moore, D. I. Skoles, and N. Myers. 1987. The flux of carbon from terrestrial ecosystems to the atmosphere in 1980 due to changes in land use: geographic distribution of the global flux. *Tellus* 39B:122–139.

Hunsaker, C. T., R. V. O'Neill, B. L. Jackson, S. P. Timmins, and et al. 1994. Sampling to characterize landscape pattern. *Landscape Ecol*ogy 9:207-226.

Jenerette, G. D., and J. Wu. Analysis and simulation of land-use change in the central Arizona–Phoenix region. *Landscape Ecology* 16:611–626.

Jones, B., J. Walker, K. H. Riitters, J. D. Wickham, and C. Nicoll. 1996. Indicators of landscape integrity. In *Indicators of catchment health: A technical perspective*, edited by J. Walker and D. J. Reuter. Melbourne: CSIRO.

Knapp, A. K., J. T. Fahnestock, S. P. Hamburg, L. B. Statland, T. R. Seastead, and D. S. Schimel. 1993. Landscape patterns in soil-plant water relations and primary production in tallgrass prairie. *Ecology* 74:549–560.

Krummel, J. R., R. H. Gardner, G. Sugihara, R. V. O'Neill, and P. R. Coleman. 1987. Landscape patterns in a disturbed environment. *Oikos* 48:321–324.

Li, H., and J. F. Reynolds. 1997. Modeling effects of spatial pattern, drought, and grazing on rates of rangeland degradation: A combined Markov and cellular automata approach. In *Scales in remote sensing and GIS*, edited by D. A. Quattrochi and M. F. Goodchild. City: Lewis Publishers.

Lowrance, R., R. A. Leonard, L. E. Asmussen, and R. L. Todd. 1985. Landscape patterns in soil-plant water relations and primary production in tallgrass prairie. *Ecology* 66:287–296.

Ludwig, J., D. Tongway, D. Freudenberger, J. Noble, and K. Hodgkinson. 1997. *Landscape ecology, function and management: Principles from Australia's rangelands*. Collingwood: CSIRO.

McDonnell, M. J., S. T. A. Pickett, P. Groffman, P. Bohlen, R. V. Pouyat, W. C. Zipperer, R. W. Parmelee, M. M. Carreiro, and K. Medley. 1997. Ecosystem processes along an urban-to-rural gradient. *Urban Ecosystems* 1:21–36.

Moss, M. R., editor. 1988. Landscape ecology and management. *Proceedings of the first symposium of the Canadian Society of Landscape Ecology and Management, May 1987*. Montreal: Polyscience.

Newell, A., and H. A. Simon. 1972. *Human problem solving*. Englewood Cliffs, N.J.: Prentice-Hall.

O'Neill, R. V., D. L. DeAngelis, J. B. Waide, and T. F. H. Allen. 1986. *A hierarchical concept of ecosystems*. Princeton: Princeton University Press.

O'Neill, R. V., K. B. Jones, K. H. Riitters, J. D. Wickham, and I. A. Goodman. 1994. *Landscape monitoring and assessment research plan*. U.S. EPA 620/R-94/009. Washington, D.C.: publisher.

O'Neill, R., C. Hunsaker, K. B. Jones, K. H. Riitters, J. D. Wickham, P. M. Schwartz, I. A. Goodman, B. L. Jackson, and W. S. Baillargeon. 1997. Monitoring environmental quality at the landscape scale. *BioScience* 47:513–519.

Ojima, D. S., K. A. Galvin, and I. Turner, B. L. 1994. The global impact of land-use change. *BioScience* 44:300–304.

Parton, W. J., J. W. B. Stewart, and C. V. Cole. 1988. Dynamics of C, N, P and S in grassland soils: A model. *Biogeochemistry* 5:109–131.

Parton, W. J., D. S. Schimel, C. V. Cole, and D. S. Ojima. 1987. Analysis of factors controlling soil organic matter levels in Great Plains grasslands. *Soil Science Society of America Journal* 51:11731179.

Parton, W. J., J. M. O. Scurlock, D. S. Ojima, T. G. Gilmanov, R. J. Scholes, D. S. Schimel, T. Kirchner, J.-C. Menaut, T. Seastedt, E. G. Moya, A. Kamnalrut, and J. I. Kinyanmario. 1993. Observations and modeling of biomass and soil organic matter dynamics for grassland biome worldwide. *Global Biogeochemical Cycles* 7:785–809.

Pickett, S., W. R. Burch, S. E. Dalton, T. W. Foresman, J. M. Grove, and R. Rowntree. 1997. A conceptual framework for the study of human ecosystems in urban areas. *Urban Ecosystems* 1:185–199.

Pickup, G. 1985. The erosion cell—A geomorphic approach to landscape classification in range assessment. *Australian Rangeland Journal* 7:114–121.

Pielke, R. A., and R. Avissar. 1990. Influence of landscape structure on local and regional climate. *Landscape Ecology* 4:133–155.

Reynolds, J. R., and J. Wu. 1999. Do landscape structural and functional units exist? In *Integrating hydrology, ecosystem dynamics, and biogeochemistry in complex landscapes,* edited by J. D. Tenhunen and P. Kabat. New York: John Wiley.

Reynolds, J. F., D. W. Hilbert, and P. R. Kemp. 1993. Scaling ecophysiology from the plant to the ecosystem: a conceptual framework. In *Scaling physiological processes: Leaf to globe,* edited by J. R. Ehleringer and C. B. Field. San Diego: Academic Press.

Reynolds, J. F., R. A. Virginia, and W. H. Schlesinger. 1997. Defining functional types for models of desertification. In *Plant functional types,* edited by T. M. Smith, H. H. Shugart, and F. I. Woodward. Cambridge: Cambridge University Press.

Risser, P. G. 1990. Landscape pattern and its effects on energy and nutrient distribution. In *Changing landscapes: An ecological perspective,* edited by I. S. Zonneveld and R. T. T. Forman. New York: Springer-Verlag.

Schimel, D. S., V. Participants, and B. B.H. 1997. Continental scale variability in ecosystem processes: models, data, and the role of disturbance. *Ecological Monographs* 67:251–271.

Shugart, H. H. 1998. *Terrestrial ecosystems in changing environments*. Cambridge: Cambridge University Press.

Simon, H. A. 1973. The organization of complex systems. In *Hierarchy theory: The challenge of complex systems,* edited by H. H. Pattee. New York: George Braziller.

———. 1962. The architecture of complexity. *Proceedings of the American Philosophical Society* 106: 467–482.

Sklar, F. H., and R. Costanza. 1991. The development of dynamic spatial models for landscape ecology. A review and prognosis . In *Quantitative methods in landscape ecology,* edited by M. G. Turner and R. H. Gardner. New York: Springer-Verlag.

Turner, M. G. 1987. Spatial simulation of landscape changes in Georgia: A comparison of three transition models. *Landscape Ecology* 1:29–36.

Turner, M. G., and R. H. Gardner. 1991. *Quantitative methods in landscape ecology: The analysis and interpretation of landscape heterogeneity.* New York: Springer-Verlag.

U.S. Census Bureau. Year. World wide web document available at www.census.gov. Viewed on date.

Vitousek, P., and H. Mooney. 1997. Human domination of earth's ecosystems. *Science* 277:494–499.

Vitousek, P. M., J. D. Aber, R. W. Howarth, G. E. Likens, P. A. Matson, D. W. Schindler, W. H. Schlesinger, and D. G. Tilman. 1997. Human alteration of the global nitrogen cycle: Sources and consequences. *Ecological Applications* 7:737–750.

Wondzell, S. M., G. L. Cunningham, and D. Bachelet. 1996. Relationships between landforms, geomorphic processes, and plant communities on a watershed in the northern Chihuahuan Desert. *Landscape Ecology* 11:351–362.

Wu, J. 2000. Urban ecology (book review). *Ecological Engineering* 16:581–582.

———. 1999. Hierarchy and scaling: Extrapolating information along a scaling ladder. *Canadian Journal of Remote Sensing* 25:367–380.

———. 1998. Simulating urban encroachment on rural land with fuzzy-logic-controlled cellular automata in a geographical information system. *Journal of Environmental Management* 53:293–308.

Wu, J., and S. A. Levin. 1994. A spatial patch dynamic modeling approach to pattern and process in an annual grassland. *Ecological Monographs* 64 (4):447–464.

Wu, J., and O. L. Loucks. 1995. From balance-of-nature to hierarchical patch dynamics: A paradigm shift in ecology. *Quarter Review of Biology* 70:439–466.

Wu, J., and Y. Qi. 2000. Dealing with scale in landscape analysis: An overview. *Geographic Information Sciences* 6:1–5.

Wu, J., D. E. Jelinski, M. Luck, and P. T. Tueller. 2000. Multiscale analysis of landscape heterogeneity: Scale variance and pattern metrics. *Geographic Information Sciences* 6:6–19.

Zipperer, W. C., J. Wu, R. V. Pouyat, and S. T. A. Pickett. 2000. The application of ecological principles to urban and urbanizing landscapes. *Ecological Applications* 10:685–688.

Adaptive Management of Complex Socio-environmental Systems in the Southwestern United States: Examples of Urbanizing Watersheds in Arizona and Texas

Laura R. Musacchio, William E. Grant, and Tarla R. Peterson

Introduction

The sustainable use of natural resources in urbanizing watersheds of the southwestern United States is one of the nation's most pressing environmental policy, planning, and management problems. For many decades, these watersheds were considered unproductive when compared to those in the eastern United States because of the lack of year-round water supply. Water diversion projects, which were subsidized by the federal and state governments, allowed year-round water supplies for agriculture and communities. Despite these projects, the watersheds remained largely rural and had natural resource–based economies that were focused on ranching, agriculture, and mining.

Rapid human population growth in the major cities such as Phoenix and San Antonio has transformed these largely rural watersheds into urbanizing watersheds that are increasingly complex as socio-environmental systems. The region faces a number of unique challenges that are related to water resources: interdependent urban and rural uses, finite water supply, flood-prone watersheds, loss of riparian habitats, and contamination of water supplies. In addition, some of the watersheds drain areas that are located in the United States and Mexico. The complexities of the water-related problems that face this region present the challenge/opportunity for adaptive management of complex socio-environmental systems.

In this paper, we first provide an overview of the systems perspective, which provides a powerful conceptual framework for dealing with complex systems, such as urbanizing watersheds. We then discuss adaptive management as an application of the systems perspective to deal with the complexity and uncertainty inherent in environmental decision-making and management. Finally, we consider adaptive management within the context of the socio-environmental complexity of two urbanizing watersheds in the southwestern United States; we present case studies of the watersheds of Phoenix's West Valley (Arizona) and the San Antonio River watershed (Texas).

The Systems Perspective

Arguably, the most critical problem that we face, whether it be as educators, scientists, resource managers, or ordinary citizens, is recognizing causal relationships in the complex systems in which we work and live (Forrester 1994). Experiences with simple systems from early childhood teach us that cause and effect are tightly linked in space and time; a hand placed too close to a flame leaves little doubt concerning cause and effect. However, in complex systems an apparent local cause often is only correlated with our original observation, both having resulted from the same temporally and spatially distant true cause. This fosters decisions that are at best ineffective, often counterproductive, and sometimes disastrous.

The systems perspective (von Bertalanffy 1968; Laszlo 1996) facilitates the recognition of causal relationships in complex systems that cannot be identified by other methods of problem solving, thus providing a context within which to view current problems/opportunities in the management of complex socio-environmental systems that is both unique and powerful. Various schemes for applying the systems approach in ecology and natural resource management have been suggested (Patten 1971; Jeffers 1978; Kitching 1983), but all are based on the same underlying general systems theory (Ashby 1956). We might identify four fundamental phases in the process of developing and using a quantitative systems simulation model: (1) conceptual model formulation; (2) quantitative model specification; (3) model evaluation; and (4) model use (Grant 1986; Grant et al. 1997). The goal of the first phase is to develop a conceptual, or qualitative, model of the system of interest. Based on objectives of the modeling project, we decide which components in the real world system should be included in our system of interest and how they should be related to one another. The goal of the second phase is to develop a quantitative model of the system of interest. This basically involves translating our conceptual model, which is represented diagrammatically and using words, into a series of mathematical equations that collectively form the quantitative model. This translation, or quantification, is based on consideration of various types of information about the real system including theoretical concepts, empirical data, and expert opinion. The goal of the third phase is to evaluate the usefulness of the model in meeting our objectives. Within a management context, often we are particularly interested in determining how sensitive model predictions are to the uncertainties with which we have represented various aspects of the model. The final phase involves designing and simulating the same experiments with the model that we would conduct in the real system to answer our questions.

The four phases of systems analysis are highly interconnected, and we usually cycle through the phases repeatedly (Grant et al. 1997). We might imagine a scenario in which we start with few data and little understanding. We first integrate existing knowledge via systems analysis and simulation—conceptual model formulation, quantitative model specification, model evaluation, and model use—to generate hypotheses about how the system works. This model-building process increases our understanding of the system and identifies specific areas in which important data are lacking. As we accumulate (by experimentation or observation in the field or laboratory), analyze (qualitatively or quantitatively, perhaps statistically), and interpret more relevant information, we generate further hypotheses concerning system structure and function that can be integrated into our simulation models. Further simulation increases understanding and identifies more information needs, and so on. As we continue to accumulate, analyze, interpret, and integrate relevant information, we increase our understanding of the system and obtain better solutions to our problem.

Obviously, this scheme lends itself well to the collaborative learning that is an integral part of adaptive management. One of the main obstacles in moving toward sustainable, multiple use of resources is our inability to synthesize knowledge and perspectives from many distinct disciplines within a single problem-solving philosophy. Another barrier is our inability to identify and communicate "high-leverage" policies to decision makers. Forrester (1994) has suggested that perhaps 98 percent of the alternatives considered for managing a system are "low-leverage" policies; that is, they have low potential to create change in system dynamics. Yet, most heated debates in communities, companies, and government are about low-leverage policies, which diverts attention from the few policies that could lead to dramatic improvement. Forrester (1994) further suggests that when working with a high-leverage policy, 90 percent of the time decision makers push the lever in the wrong direction because they rely on intuition rather than on an understanding of system function.

The systems perspective provides a basis for sustainable, multiple resource use by facilitating multidisciplinary planning and by creating an effective communication interface between scientists, citizens, and policy makers, thus promoting identification and communication of high- versus low-leverage policies to decision makers (Grant 1998). The concept of using the systems perspective as the basis for resource management has been articulated eloquently in books such as *Adaptive Environmental Assessment and Management* (Holling 1978) and *Adaptive Management of Renewable Resources* (Walters 1986). Practical application of the concept often involves a series of workshops in which a core team of disciplinary specialists, systems modelers, citizens, and decision makers focuses on development of a simulation model to address specific management questions (Grant 1998). The simulation model evolves into the communication interface between scientists and decision makers as team members develop a sense of joint ownership of the model, thus making the modeling process more important than the model itself.

Adaptive Management

Adaptive management emphasizes how policies can be designed to cope with uncertainty in environmental decision-making and management (Lee 1993). With this approach, policy making is viewed as a hypothesis-driven experiment that is informed by learning (Lee 1993). Adaptive management evolved from the research of C. S. Holling and the Institute of Applied Systems Analysis in Austria. In the 1970s, Holling called the approach *adaptive environmental assessment*, which merged the theory of systems ecology and natural resource management. The approach eventually came to be known as *adaptive management*. One of the most important assumptions of the approach is that understanding the behavior of the system of interest is essential for designing better policies (Holling 1978). The importance of the systems perspective is stressed, thus ecological, social, economic, and cultural factors are considered early in environmental decision-making and the management process (Lee 1993). In addition, constant interaction between scientists, managers, and other stakeholders is encouraged because this feedback is seen as a form of social learning and improves the policy making process (Christensen et al. 1996; Lessard 1998). Computer models, especially systems simulation models, are an important technique for comparing the potential effects of policies versus expected effects and for including the input of stakeholders into the decision-making process (Holling 1978).

Scientists and managers have applied adaptive management to the Everglades, Chesapeake Bay, and Columbia River Basin, which have significant water-related problems. At the heart of each of these examples is a public resource of national significance that is affected by the control of an ecological variable in order to produce some good for human consumption such as timber, electricity, water, or food. In the past, these ecological variables were managed in such a way that the resiliency of the ecosystems was compromised. Holling (1995) described the characteristics of management policies that contributed to the loss of ecosystem resiliency such as a single target and piecemeal policy with a single scale of focus. Adaptive management experiments have brought an entirely different perspective to policy making. Integrated policies with a long-term perspective are the rule. One of the challenges of adaptive management is the number of people and cost of the experiments. If these case studies are successful, they will be important benchmarks for urbanizing watersheds in the southwestern United States.

Adaptive management has not been actively applied to watersheds that are dominated by urban land. Application in such ecosystems could be significant because of the complexity of public policy and ecosystem resiliency issues. However, this management approach needs refinement because urbanizing land is not a renewable natural resource, which is the typical situation in other case studies of adaptive management. In addition, adaptive management would need to address specific types of socio-environmental complexity such as more stakeholders, landowners, and institutions. However, adaptive management is designed to be a collaborative, iterative, and experimental process so in many ways it is ideally suited for application to urbanizing watersheds because urban planning employs a similar process. In addition, potential exists for the linkage between science, policy, monitoring, planning, and management.

One of the definitive characteristics of urbanizing watersheds is the rate and complexity of change that is occurring in the socio-environmental systems. For example, in the southwestern United States the urban edge of Phoenix expanded by one-half mile each year from the 1970s to the 1990s (Morrison Institute for Public Policy 2000). During the same time, urbanization converted open space around the urban fringe of Phoenix: agricultural land decreased by 40 percent and undeveloped open desert decreased by 32 percent (Morrison Institute for Public Policy 2000).

The rapid rate of urbanization in watersheds of the southwestern United States has significant impacts on water quality and water quantity. Flooding, groundwater contamination, and non-point source (NPS) pollution are just a few of the problems that not only impact cities but also communities located downstream. Traditional management strategies can fail in these situations because of the systemic complexity of the environmental issues in watersheds. Therefore, the scientists, managers, and planners that are engaged in these issues have become more open to alternative management approaches.

System simulation models that are used in adaptive management and urban planning could be important tools in land use decision making, policy, and management within the complex socio-environmental milieu of urbanizing watersheds. System simulation models could communicate the linkage between decision making, environmental impacts, and land use planning. This type of model could demonstrate how different intensities of land use could impact the flooding regime of a river. A group of researchers from Texas A&M University is using such a model with citizens who reside in the San Antonio River watershed. Another approach is the integration of system simulation models, geographic information systems (GIS), and computer-aided design (CAD) that is proposed for the Agua Fria River watershed in Arizona. Such an approach can be used in land use planning to aid the visualization of alternative schemes for river restoration and ecological processes at the watershed and floodplain scales.

Socio-environmental Complexity in Urbanizing Watersheds of the Southwestern United States

We now explore more specifically how adaptive management could be applied to the watersheds of Phoenix's West Valley in Arizona and the San Antonio River watershed in Texas. For each watershed, we review the major water resource issues and socio-environmental conditions, the major institutions and stakeholders involved in water resource management, and the role of computer models in the dynamics of the decision-making process.

Overview

The cities of Phoenix and San Antonio represent of a new form of American urbanism typical of the Sunbelt (Fink 1993). Postwar public policy influenced the growth patterns of these cities by providing funding for new highways and tax incentives for single-family homeownership (Fink 1993). Consumption of land increased and population density decreased as the single-family home and suburbia became the benchmark for success. New urban forms emerged, such as edge cities. The cities of the southwestern United States epitomize these new urban growth patterns. In addition, these cities are influenced by many of the same biophysical and sociocultural trends.

The watersheds of Phoenix's West Valley and San Antonio River watershed have been impacted by the rapid growth of the postwar years. However, public policy has not typically emphasized the importance of urbanization on water resources. The Environmental Protection Agency's Clean Water Action Plan (1998) represents one of the most important benchmarks in the recognition of this environmental problem. In addition, the Environmental Protection Agency (EPA) developed indices of condition and vulnerability for all watersheds in the United States. These indices represent an important beginning point for the study of the watersheds of Phoenix's West Valley and San Antonio River watershed. The West Valley includes three watersheds: Lower Salt, Gila, and Agua Fria. Despite being the most urbanized watershed, the Lower Salt has achieved the rating of better water quality, but it has high vulnerability for future degradation from urban growth. On the other hand, the Gila River has less serious water quality issues with low vulnerability because most of the watershed is farm and desertlands. Because the Agua Fria River watershed has been on the suburban fringe of Phoenix until recently, it has the highest EPA indicator rating, which is given to rivers of better quality with low vulnerability (EPA 2000). In comparison, the San Antonio River watershed has a moderate EPA indicator rating, which identifies less serious water problems with low vulnerability.

The watersheds of Phoenix's West Valley will be one of the most important areas for growth in the Phoenix metropolitan area in the next 20 years. Agriculture and suburban land are the most common land uses in these watersheds. Affordable land and close proximity to downtown Phoenix will make these watersheds attractive for the rapid expansion of suburban development. Because of the proposed suburban growth, potential watershed vulnerabilities include human population growth, aquatic species at risk, loss of riparian habitat, and urban runoff potential. Hydrological modifications include the Waddell Dam, Lake Pleasant, regional canal system, sand and gravel mines, and river channelization. However, mitigation of past hydrological modifications through river restoration projects is an important trend in these watersheds. The Tres Rios Constructed Wetlands Demonstration Project is an excellent example of this project type.

A complex web of stakeholders influences the socio-environmental conditions in the watersheds of Phoenix's West Valley. The communities along the river are ethnically, socially, and economically diverse. For example, the city of Goodyear was established as a small company town founded by the Goodyear Rubber Company and is now a rapidly growing suburb. Another example is the city of El Mirage is one of the area's oldest Hispanic communities. The residents of these communities have their own vision of how the rivers will improve their quality of life in the face of rapid suburban growth. Yet, the challenge for planners at such agencies as the Maricopa County Flood Control District is to address the concerns of each community in the water course and recreation master plans while also addressing issues such as flooding, erosion, riparian habitat, sand and gravel mining, and water quality.

The San Antonio River watershed typifies several natural resource management concerns that are becoming pervasive throughout the American Southwest. For example, human population growth has a serious impact on quality and quantity of freshwater resources. San Antonio is the only major city in the Southern Edwards region of Texas, where the ratio of agricultural to municipal water use was 50:40 in 1990, but by 2030 is predicted to be 30:60. Declining levels of aquifers, increased nutrient and contaminant loading of streams, decreasing freshwater inflow to estuaries, loss of endemic species, and invasion of exotic species represent human-induced ecological problems with serious socioeconomic repercussions. Municipal water use accounts for 71 percent of water used in the San Antonio River Basin (Texas Water Development Board 1997).

Rehabilitation of this urban watershed will require a significant effort from stakeholders in the area. The social heterogeneity of the area presents additional complications. In addition to the economic diversity of the watershed, there is also diversity in ethnic background and income level. According to the recent state counts, Bexar County, through which the San Antonio River runs, was 53.9 percent Hispanic, 37.5 percent Anglo, and 6.6 percent Black in 1995. The fastest growing segment of the population is located in the "Other" category and is made up primarily of Asians. Although trade, government, and services dominate employment; construction, manufacturing, utilities, mining, and agriculture can also be found in the county (Controller's Office 1995). This watershed, therefore, represents the kind of mix of cultural groups and economic interests that tend to characterize urban areas.

The watersheds of Phoenix's West Valley

The watersheds of Phoenix's West Valley are some of the best examples of urban fringe watersheds that will experience rapid land use and land cover change in the next decade. The communities along these rivers have different plans for the management of these resources. Some communities want to restore riparian habitat while others see the river as a source of economic renewal, and still others allow sand and gravel mines to excavate the riverbed. Floodplain managers and urban planners have created several master plans that help to integrate the multiple visions for the Salt, Gila, and Agua Fria Rivers. The common denominator of all the plans is the acknowledgment of the river as an important public resource for multiple uses. The plans are influenced by what Kellert (1996) describes as different human values toward an ecological systems. However, there are some nagging questions about how these plans integrate multiple uses and address potential changes to the river system as an ecosystem. All the plans seek to control a target variable, the ecological state of the river system, but for different purposes. Yet, ecological change is not an explicit question that is answered during the planning process or in the plan. The question is whether all of these plans create ecological change in the river system that makes it less resilient and thus more vulnerable.

The purpose of this example is to outline the current opportunities and constraints for the application of adaptive management to the watercourse and restoration planning process in the watersheds of Phoenix's West Valley. The primary focus is how to address two critical issues, avian habitat and water resources, that are both public resources but are often located on private land, particularly the riparian and floodplain zones of the river. Because of these issues, adaptive management could potentially be used as the framework for a regional approach to these issues. However, the adaptive management will need to be integrated with theory from landscape ecology, landscape planning, urban planning, knowledge engineering, and watershed management in order to address avian habitat and water resource issues in this rapidly urbanizing watershed. The challenge is to link the modeling of the ecological state of the river to the state of human decision making. In this way, the model can address how the dynamics of socio-environmental variables and human decision making causes change in the land use and habitat patterns of the watersheds of Phoenix's West Valley.

Given these considerations, how could urban planners, civil engineers, floodplain managers, and policy makers use adaptive management to address avian habitat and water resources in watercourse master plans and river management plans in urbanizing watersheds such as that of the watersheds of Phoenix's West Valley? The challenge is to find integrated regional landscape modeling approaches that can represent how land-use decision making influences the types of conservation, preservation, and restoration strategies at the watershed, floodplain, and riparian scales. A recent report by Dale et al. (2000) state that approaches applied to federal land are the best examples of the incorporation of ecological principles into land-use decision making. A number of these examples used adaptive management as the basis for environmental problem solving. Adaptive management could be used in the watersheds of Phoenix's West Valley because federal and state agencies own large tracts of land in these watersheds. However, private land is most common in these watersheds, so adaptive management would have to be integrated with existing land use planning processes, such as the watercourse master plan. A model that compares land-use decision making on public versus private land in a watershed would be challenging to develop but highly informative.

The modeling approach for the watersheds of Phoenix's West Valley will build on the study by Musacchio (1999). This approach used a system simulation model and GIS model to link landscape ecology, land use planning, and adaptive management. Its strength is its conceptual foundation in systems theory, which addresses socio-environmental complexity of the rural landscape. However, this approach is focused on agricultural policy. In order to adapt this approach to land use planning in urbanizing watersheds, the modeling approach needs to be modified to fit the format of public workshops that are used in the watercourse master planning process. The modeling approach will include the use of system simulation models and GIS and CAD. The system simulation models with such software as STELLA 6.0 (High Performance Systems, Inc.), are one of the most flexible tools because a modeler can use the modeling process described in the second section of this paper to link human decision making to socio-environmental changes in urbanizing watersheds. GIS software, such as Spatial Analyst in ArcGIS, allows for changes in urbanizing watersheds to be linked to spatial patterns such as land use and land cover. CAD software, such as Form-Z, Rapidsite, and 3-D Studio, is often underutilized in the planning process because this technology is associated with the design professions. However, CAD, GIS, and STELLA“ 6.0 could be linked via Visual Basic for a more sophisticated approach to land-use decision making, allowing for four-dimensional representations of change in complex socio-environmental systems in the Agua Fria River watershed. The goal would be a model that could potentially allow the input of data from the long-term monitoring of the Central Arizona–Phoenix Long-Term Ecological Research Project.

San Antonio River watershed

The San Antonio River Basin traverses at least three ecoregions in Texas: the Central Texas Plateau, the Texas Blackland Prairies, and the Western Gulf Coastal Plain. This basin is dominated by urban and industrial development from the City of San Antonio, Texas, but agriculture is also a major economic force in the counties through which the basin runs. Water quality in the basin is declining due to NPS pollution from the urban environment (TNRCC 1996). Segments of this watershed are unsafe for human or wildlife use due to elevated concentrations of toxic metals, fecal coliform bacteria, and nutrients. These pollutants are believed to derive from urban NPS runoff (TNRCC 1996). Water quality in these river reaches must be restored by 2003 under the Texas Natural Resource Conservation Commission's (TNRCC) Statewide Basin Management Schedule (TNRCC 1997). Rehabilitation of the water bodies is required under the Environmental Protection Agency's Total Maximum Daily Load (TMDL) strategy for watershed restoration.

The sources of pollutants in this urban watershed are diverse and complex. Their rehabilitation will require commitment and agreement among the varied stakeholders in them. Top-down autocratic resource management methods have not proven effective in protecting these valuable ecological services. A process model of the watershed could provide a means for dealing with the varied perspectives and interests of stakeholders. Through collaborative interaction between scientists and stakeholders, we are developing a system simulation model that incorporates salient water quantity and quality issues. It incorporates iterative input from stakeholders to guide risk-based research and restoration planning.

Two stakeholder groups have been convened with the assistance of the San Antonio River Authority (SARA). Members have been trained in the methods and goals of collaborative learning (CL) (Daniels and Walker 1995), a process that has been used successfully with other large, heterogeneous policymaking groups. The CL process intervention (Daniels and Walker 1995, 1996; Daniels et al. 1996) is grounded in theoretical work on soft systems methodology (Checkland 1981; Checkland and Scoles 1990; Flood and Jackson 1991; Wilson and Morren 1990) and alternative dispute resolution (Fisher et al. 1991; Gray 1989; Moore 1986; Susskind and Cruikshank 1987).

The concept of *soft systems* represents an extension of theoretical work on systems analysis and experiential learning. The basic assumption in soft systems methodology is that management of complex problem situations ("fuzzy" or ill-defined problems) demands a focus on process. Instead of the hard systems emphasis on problem-solution, the soft systems approach focuses on "situation improvements" that can result from active learning and debate. Thus, soft systems methodology stresses the need for systems thinking (Senge 1990) and the importance of experiential learning (Kolb 1986). Checkland and Scoles (1990) argue that soft systems methodology is best suited for situations in which the problem itself is ambiguous and subject to different interpretations. Management of ecosystems, particularly the issue of NPS pollution and decision making about TMDLs is precisely the type of situation where problems are often characterized by a high degree of uncertainty and equivocality. The stakeholder groups are using this model to assist the development of specific recommendations for watershed restoration. They are using the learning achieved through developing and manipulating these models to work through how their action plans could be implemented on the ground in their respective areas of the watershed. For example, they have used the model to better understand and communicate how different intensities of land use would interact with various channel conditions and retention structures to impact the watershed's flooding regime.

Discussion

The case studies of the watersheds of Phoenix's West Valley and San Antonio River watersheds represent the shift in environmental management and urban planning from top-down approaches with teams composed solely of experts to bottom-up approaches with teams of experts, citizens, and other stakeholders. Through the use of computer models in workshop settings, adaptive management facilitates a process that encourages systemic analysis of environmental problems by stakeholders. This collaborative process is a hybrid between the hard and soft systems approaches. The hybrid nature of the process addresses the feedbacks between the human alteration of land use and land cover and shifts in the hydrological and ecological processes of urbanizing watersheds. Such an approach is needed in urban planning since this discipline generally addresses social and economic concerns. Ecological change in urban systems is rarely addressed in urban planning.

The use of the hybrid approach responds to the rapid social and environmental changes in the urbanizing watersheds of the southwestern United States. The rapid growth of the human population since World War II and the type of urban settlement patterns have left the region's water resources in a state of overdraft. Difficult decisions await planners, scientists, and citizens if a serious regionwide drought occurs. Recent electrical shortages in California and the limited capacity of existing hydroelectric plants to meet these demands are indicative of the potential impact of limited water resources on the allocation of water for urban, agricultural, and environmental uses in this region.

Both case studies represent the first step toward the use of integrated land use models in the urbanizing watersheds in the southwestern United States. The case studies illustrate two potential applications of different computer models. The example of the San Antonio River uses a system simulation model that couples a hydrological model with a decision-making model. The graphic interface of the model encourages intuitive exploration of the link between different types of human decision making, hydrological regimes, and engineering solutions. In contrast, the study focused on the watersheds of Phoenix's West Valley will link different types of software to aid the three-dimensional visualization of human decision making and change in land use and land cover of riparian and floodplain zones. The models will address functional and spatial change over time.

The use of the systems perspective in the models allows multiple hypotheses to be tested by the stakeholders. The process is unusual because it blends the scientific method with collaborative planning. The iterative nature of the model building process has much in common with the iterative nature of plan making. Thus, these models have the potential to link the plan making process with community stakeholders. Integrated policies with a long-term perspective are one of the benefits, but the other benefits include plans that are more grounded in the unique decision making process of each group of stakeholders. This results in dynamic plans that address the functional and structural changes in rapidly urbanizing watersheds.

Conclusion

At a time of unprecedented population growth, the future of the rapidly urbanizing watersheds in the southwestern United States rests in the decision making of communities. The development of a public consensus on many watershed related issues, such as NPS pollution, water use, and flooding, is a challenge. The goal is the development of high-leverage policies that improve environmental management problems in the watersheds. Yet, low-leverage policies often consume most of the energy of governmental officials, citizens, planners, and scientists. The real problems are not addressed because of the systemic nature of many environmental problems in rapidly urbanizing watersheds. Adaptive management, as discussed in this paper, is one of the potential approaches for the development of more integrated environmental problem solving in communities. Through the use of computer models in public workshops, stakeholders can participate more actively in social learning and thus visualize the links between causes, effects, and decisions. Both case studies offer alternative ideas for the use of such technology, which is influenced by the particular environmental problems of the watersheds and the decision making process that is used in the communities. Hopefully, they will help promote the application of adaptive management to environmental problems in the urban environment.

References

Checkland, P., 1981. *Systems thinking, systems practice*. New York: Wiley.

Checkland, P., and J. Scholes. 1990. *Soft systems methodology in action*. New York: Wiley.

Christensen, N. H., A. M. Bartuska, J. H. Brown, S. Carpenter, C. D'Antonio, R. Francis, J. F. Franklin, J. A. MacMahon, R. F. Noss, H. Charles, M. G. Turner, and R. G. Woodmansee. 1996. The report of the Ecological Society of America Committee on the scientific basis for ecosystem management. *Ecological Applications* 6:665–691.

Dale, V. H., S. Brown, R. A. Haeuber, N. T. Hobbs, N. Huntly, R. J. Naiman, W. E. Riebsame, M. G. Turner, T. J. and Valone. 2000. Ecological principles and guidelines for managing the use of land. *Ecological Applications* 10:639–670.

Daniels, S. E., and G. B. Walker. 1995. Searching for effective natural-resources policy: The special challenges of ecosystem management. In *Ecosystem management of natural resources in the Intermountain West,* edited by F.H. Wagner. Logan: College of Natural Resources, Utah State University.

———. 1996. Collaborative learning: Improving public deliberation in ecosystem-based management. *Environmental Impact Assessment Review* 16:71–102.

———. In press. *Working through environmental conflicts: The collaborative learning approach*. Westport, Conn.: Greenwood Publishing Group.

Environmental Protection Agency. 1998. *Clean water action plan: Restoring and protecting America's waters*. Washington, D.C.: U.S. Government Printing Office.

———. 2000. *Surf your watershed*. Web document available at http://www.epa.gov/surf3/locate/. Accessed on April 15, 2001.

Fink, M.,1993. Toward a Sunbelt urban design manifesto. *Journal of the American Planning Association* 59:320–333.

Fisher, R., W. Ury, and B. Patton. 1991. *Getting to YES*. 2nd ed. New York: Penguin.

Flood, R. L., and M. C. Jackson. 1991. *Creative problem solving: Total systems intervention*. New York: Wiley.

Forman, Richard T. T. 1995. *Land mosaics: The ecology of landscapes and regions*. Cambridge: Cambridge University Press.

Forrester, J. W. 1994. Learning through system dynamics as preparation for the 21st century. *Creat. Learn. Exch.* 3, no. 3:1–8.

Gray, B. 1989. *Collaborating: Finding common ground for multiparty problems*. San Francisco: Jossey-Bass.

Grant, W. E. 1986. *Systems analysis and simulation in wildlife and fisheries sciences*. New York: Wiley.

———. 1998. Ecology and natural resource management: Reflections from a systems perspective. *Ecological Modelling* 108:67–76.

Grant, W. E., E. K. Pedersen, and S. L. Marín. 1997. *Ecology and natural resource management: Systems analysis and simulation.* New York: Wiley.

Holling, C. S. 1978. *Adaptive environmental assessment and management.* Chichester, U.K.: John Wiley & Sons.

———. 1995. What barriers? What bridges? In *Barriers and bridges to the renewal of ecosystems and institutions,* edited by L. H. Gunderson, C. S. Holling, and S. S. Light. New York: Columbia University Press.

Jeffers, J. N. R. 1978. *An introduction to systems analysis: With ecological applications.* Baltimore, Md.: University Park Press.

Kellert, S. R. 1996. *The value of life: Biological diversity and human society.* Washington, D.C.: Island Press.

Kitching, R. L. 1983. *Systems ecology: An introduction to ecological modelling.* St. Lucia, Australia: University of Queensland Press.

Kolb, D. A. 1986. *Experiential learning: Experience as the source of learning and development.* Englewood Cliffs, N.J.: Prentice-Hall.

Laszlo, E. 1996. *A systems view of the world: A holistic vision for our time.* Cresskill, N. J.: Hampton Press.

Lee, K. N. 1993. *Compass and gyroscope: Integrating science and politics for the environment.* Washington, D.C.: Island Press.

Lessard, G. 1998. An adaptive approach to planning and decision-making. Landscape and *Urban Planning* 40:81–87.

Moore, C. 1986. *The mediation process.* San Francisco: Jossey-Bass.

Morrison Institute for Public Policy. 2000. *Hits and misses: Fast growth in metropolitan Phoenix.* Tempe: School of Public Affairs, College of Public Programs, Arizona State University.

Musacchio, L. R. 1999. A landscape ecological process for wetland, waterfowl, and farmland conservation. Ph.D. diss., Texas A&M University.

Patten, B. C. 1971. A primer for ecological modeling and simulation with analog and digital computers. In *Systems analysis and simulation in ecology,* edited by B. C. Patten. New York: Academic Press.

Senge, P. 1990. *The fifth discipline: The art and practice of the learning organization.* New York: Currency Doubleday.

Susskind, L., and J. Cruikshank. 1987. *Breaking the impasse: Consensual approaches to resolving public disputes.* New York: Basic Books.

Texas Natural Resources Conservation Commission. 1997. *Statewide Basin Management Schedule.* Austin: TNRCC.

Texas Water Development Board. 1997. *Water for Texas: A consensus-based update to the state water plan.* Austin: TWDB.

von Bertalanffy, L. 1968. *General systems theory: Foundations, development, applications.* New York: George Braziller.

Walters, C. W. 1986. *Adaptive management of renewable resources.* New York: Macmillan.

Wilson, K., and G. E. B. Morren. 1990. *Systems approaches for improvements in agriculture and resource management.* New York: Macmillan.

Dynamic Spatial Modeling of Urban Growth on the San Pedro Watershed

Gary Whysong, Tasila Banda-Sakala, and Betsy Conklin

Introduction

The San Pedro Watershed is ecologically diverse, containing desert shrub-steppe, riparian, desert grassland, oak savannah, and ponderosa pine communities. In addition, it contains a great number and diversity of mammalian species in addition to man.

From a socioeconomic perspective, great concern exists regarding the long-term viability of San Pedro riparian communities and ranching in the face of continued population growth. Groundwater sustains the riparian system on the United States side of the border and also much of the ranching industry in Mexico's portion of the San Pedro. The threat of excessive groundwater pumping, as population numbers increase, on the riparian system has prompted the first application of international environmental law within the United States, the North American Free Trade Agreement.

This study is concerned with the use of dynamic spatial modeling techniques as discussed in Brady and Whysong (1999) to spatially describe and simulate urban population growth in a portion of the San Pedro Watershed. The most popular way to develop and simulate spatial dynamic models is through the use of cellular automaton with the integration of GIS. Applications of this approach consist of urban growth modeling (Batty et al. 1989; Batty 1991; Batty and Xie 1994; Kirtland et al. 1994; Batty and Longley 1994), land-use change patterns (White and Engelen 1993, 1994), and rural residential settlement change (Deadman et al. 1993).

The objectives of this study are:

- To develop and test land-use conversion models that predict the rate and spatial pattern of urban development.
- To attempt to predict the future spatial dynamics of major urban development in the San Pedro watershed through 2010.

Possible benefits of this research may include information that will help determine or allow forecasting of future groundwater demands in this rapid urban growth region. Continued and increasing groundwater extraction is likely to have undesirable effects on riparian communities as well as on agricultural activity.

The work reported herein focuses on describing the spatial urbanization of Sierra Vista, Arizona, from 1986 to 2010 using dynamic spatial modeling techniques. Sierra Vista began with the establishment of Camp Huachuca in 1877. At that time, what came to be known as Fort Huachuca was a military outpost to protect settlers and the southern border of the United States. A small community began to grow east of Fort Huachuca, which became the present-day Sierra Vista, and has grown to cover 26.9 km2. From 1980 to 1990 Sierra Vista demonstrated a 32 percent increase in population, from 24,937 to 32,983 (ADES 1999), and is currently estimated at over 40,000.

Methods

The information available from public and non-public sources that was within project means to acquire consisted of population data, historical information, aerial imagery, and satellite imagery. Relevant GIS layers obtained from the Arizona Land Resource Information System (ALRIS) included land ownership, transportation, and slope. City boundaries were digitized from Sierra Vista city maps, which provided information concerning historical annexation over the time period covered by the study. Landsat imagery was acquired for 1986, 1992, 1995, and 1998. It was decided that 1986 would comprise the initial conditions for the model and that population and spatial information concerning urban expansion from the remaining years would be used to test model output.

Spatial resolution for the model was maintained at the 28.5 by 28.5 m resolution provided by Landsat imagery. Most prior work with modeling of this type has been conducted at significantly lower spatial resolutions. Although model accuracy is more difficult to achieve at higher spatial resolutions, the results are potentially more useful.

The model considers interactions in terms of land-use type and the influence that urbanized areas have on undeveloped cells (pixels) around them. Like Hall et al. (1995), we applied the principle of adjacency because the type of land use found in an adjacent cell influences how the next cell should be classified.

The demand for land conversion to urban use is considered driven by population growth. Thus, a simple population submodel was constructed to simulate population growth for Sierra Vista (Figure 1). A simple growth rate was hypothesized to simulate population increase. Demand for urban land was estimated as land use per capita. These parameters were estimated for 1986 and remained constant throughout all simulations. Urban land-use per capita was estimated at 1,469.4 m2 per individual. This estimate takes into account all transportation, private, commercial, and public service areas within the city limits needed to support the population.

The model was written in C++. It attempts to satisfy urban demand by scanning and evaluating all locations (cells) within the city limits during each time step. First, land suitability is evaluated using supporting GIS layers. If the cell is on a slope of greater than 15 percent, already urbanized, or non-privately owned, it is considered unavailable. Available pixels are further evaluated by scanning adjacent cells and estimating the probability of future development based on land use in close proximity. Those locations receiving high probabilities are selected and converted to urban, which reduces the demand for additional land development. Changes in city boundaries are periodically loaded during each simulation based upon historical annexation from 1986 to 1998. City limits are not changed after the last 1998 annexation for simulations providing projections to 2010.

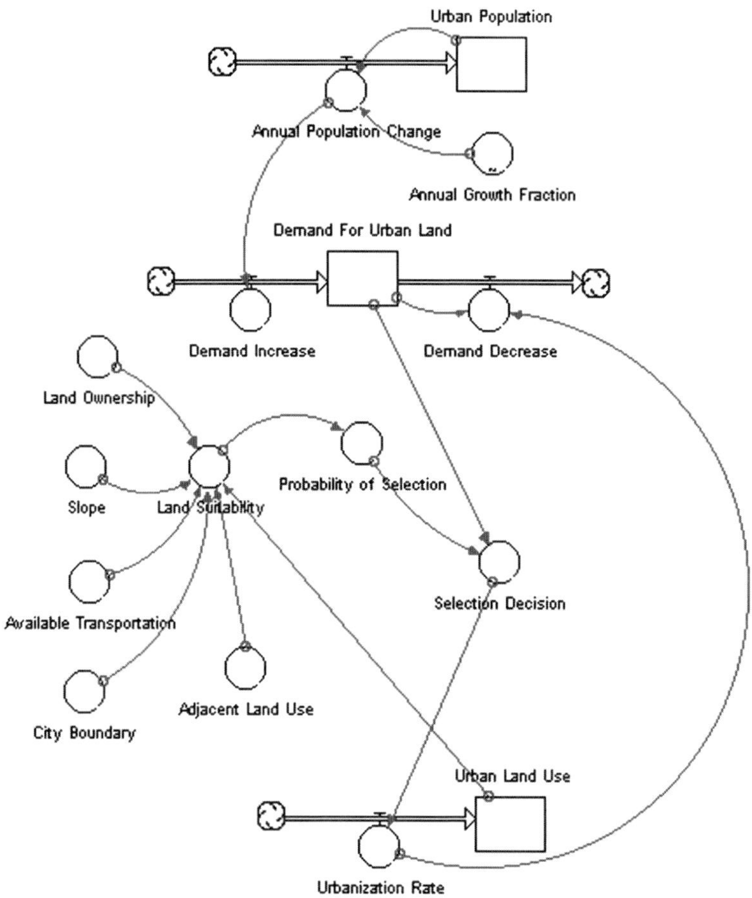

Figure 1. Conceptual flowchart of the dynamic spatial model describing land-use change for Sierra Vista, Arizona.

The additional Landsat imagery of Sierra Vista acquired for 1992, 1995, and 1998 was used to help test model accuracy. The developed urban areas within Sierra Vista's city limits were classified using supervised classification techniques and GIS layers were constructed. Progressive urban expansion was apparent and provided the basis for comparing model output. Comparisons of model accuracy were made at several spatial resolutions to investigate at which resolution, if any, that the model provided accurate results. In addition, population predictions from the population submodel were tested against Arizona Department of Economic Security population projections for agreement. Chi-square analysis was used for all statistical comparisons.

Results and Discussion

Remote sensing classification of urban areas from 1986, 1992, 1995, and 1998 Landsat imagery resulted in GIS layers having accuracies at or very near 87 percent. Although these accuracies were lower than we preferred, they were deemed sufficient to account for initial conditions and provide a basis for parameter estimation (1986 imagery only). All classified imagery for subsequent years was used for model testing.

Model simulation resulted in population growth of 26 percent from the 1986 initial conditions of 29,764 to 37,709 in 1998. By 2010 the population increased by 60 percent to 47,777. The urbanized area expanded by 111 percent from 1986 (Figure 2) to 2010 (Figure 3).

Figure 2. Composite Landsat image of Sierra Vista area in 1986.

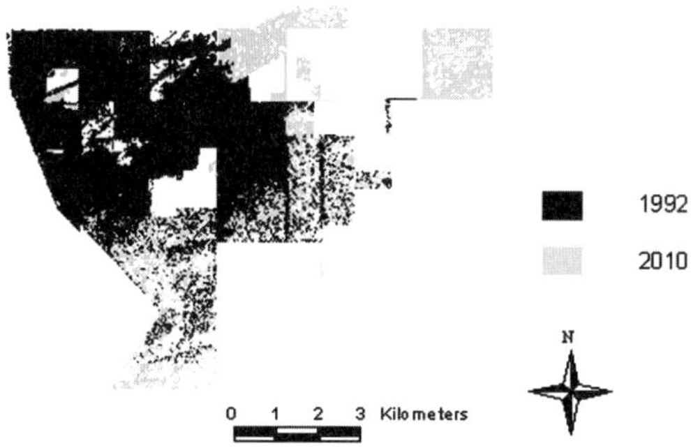

Figure 3. Urban growth of Sierra Vista illustrating expansion from 1992 simulation results to 2010.

The urban expansion resulted in an average predicted annual urban development rate of about 79 ha per year. However, the urbanization rate was not nearly as smooth as the average might imply. Several conditions present during the simulation caused the model to have greater difficulty meeting demand during some time periods (Figure 4). First, a good portion of the area within the Sierra Vista city limits is state-owned land. While state-owned land does get developed, the mechanism of land acquisition for development purposes is different than for privately owned land. No mechanism for conversion of state land was included in the model. Thus, only privately owned land was considered available for development. During certain periods, 1992 and 1996–97, development declined due to a shortage of private land within the city limits. This is evident by the strong departure of demand and developed area during those years shown in Figure 4. Acquisition of new lands through annexation by the city during these periods provided suitable new private sources for the model to attempt to meet demand. This pattern is repeated when private land becomes limited during the last few years of the simulation. No additional city boundary expansion is considered following that which occurred in 1998.

The first model tested was the population submodel. As one might expect, the population submodel could not accurately predict population numbers to within one individual ($P < 0.001$, $df = 20$). However, the model was able to predict population to within 100 individuals ($P > 0.99$, $df = 20$). This was expected, since the model was forced to describe the projected population growth data.

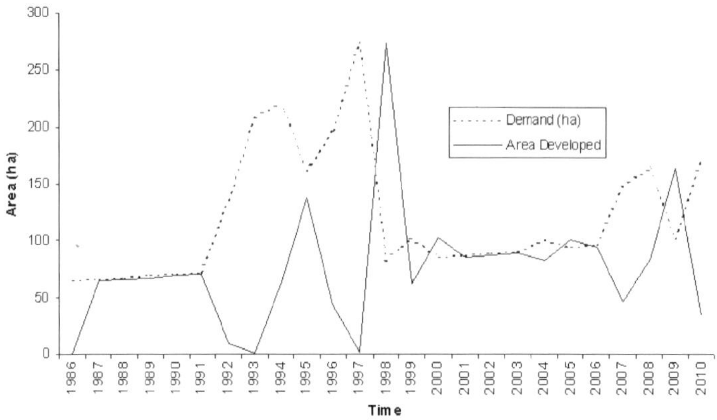

Figure 4. Simulated land-use conversion for Sierra Vista from 1986 to 2010.

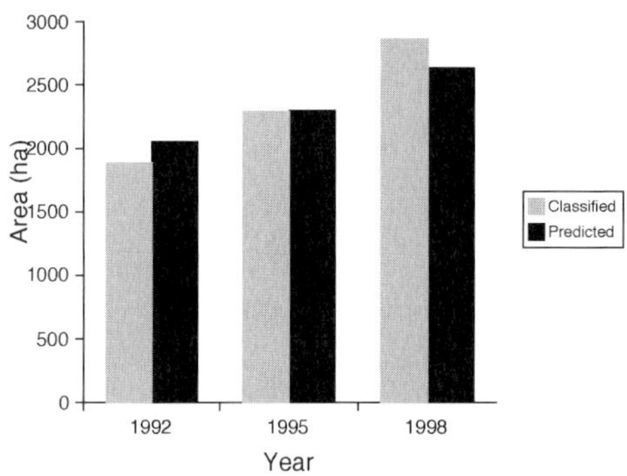

Figure 5. Classified and predicted spatial area of Sierra Vista for selected years.

The spatial model was never expected to accurately predict which areas (specific cells or pixels) would be developed in any particular time frame. This was considered impossible, since there was no means of predicting which lands would be acquired for urban development or when. Thus, the model predicted those areas "most likely" to be developed. Spatial accuracy could be assessed on the basis of total land area used for urban expansion. The classified urban land area from the 1992, 1995, and 1998 Landsat imagery was used to designate the actual changes in urban land use. This was compared to results obtained from the classified areas provided by the simulation model for the same years using a Chi-square analysis.

The urban growth trend in Sierra Vista for the years when spatial accuracy was tested is illustrated in Figure 5 at a spatial resolution of one hectare.

The tests for spatial accuracy indicated that at a resolution of one ha the model was not spatially accurate ($P < 0.03$, $df = 2$). This was not unexpected, since one ha can be considered a fairly high spatial resolution for modeling purposes. The population submodel, which drives the spatial model, can only predict population to within 100 individuals. Taking into account the per capita land use requirement (1,469.4 m2), this translates to an error of about 14.7 ha based on population estimates alone. When analyzed at a lower resolution of 0.25 km2, the difference in spatial area was not significant ($P < 0.52$, $df = 2$). Thus, the model would appear to predict the spatial conversion of land to urban use to within about 0.25 km2 accuracy.

Conclusion

This project is part of a continuing effort to model urban growth of incorporated towns over a portion of the San Pedro Watershed. The major problem, as with many modeling projects, is obtaining sufficient data to estimate model parameters and for model testing. In our case, that not only includes restricted population data, but also limited and sometimes outdated spatial information. We found the classification of remote-sensed imagery to determine urban land use very time consuming. The cause, in part, was due to a good deal of geological reflectance in the imagery that matched urban reflectance.

We do not consider the classification accuracy of the urban area for those years used to test the model output as optimal. Currently, our objectives are to obtain greater than 90 percent classification accuracy. Consequently, we are spending most of our resources on classifying remote-sensed imagery and accuracy assessment rather than modeling.

Despite the limitations, we consider this first attempt at a dynamic spatial model of urban growth in Sierra Vista a success. The spatial resolution that our model uses (28.5 m) is not common. Much work is done at lower spatial resolutions where it is easier to obtain spatial accuracy. This is apparent when it is recognized that we could not obtain a spatial accuracy for urban development at one ha, but could at 0.25 km2.

References

Arizona Department of Economic Security (ADES). 1999. Division of employee services and support: research administration—population statistics. Available at http://www.de.state.az.us/links/economic/webpage/page2.html

Batty, M. 1991. Generating urban forms from diffusive growth. *Environmental and Planning A* 23:511–544.

Batty, M., and P. Longley. 1994. *Fractal cities*. London: Academic Press.

Batty, M., P. Longley, and S. Fotheringham. 1989. Urban growth and form: Scaling, fractal geometry, and diffusion-limited aggregation. *Environmental Planning A* 21:1447–1472.

Batty, M., and Y. Xie. 1994. From cells to cities. *Environmental and Planning B: Planning and Design* 21:s31–s48.

Brady, W. W., and G. L. Whysong. 1999. Modeling. In *GIS solutions in natural resource management: balancing the technical-political equation,* edited by Stan Morain. Santa Fe: OnWord Press.

Deadman, P., R. D. Brown, and R. H. Gimblett. 1993. Modeling rural residential settlement patterns with cellular automata. *Journal of Environmental Management* 37:147–160.

Hall, C. A. S., H. Tian, Y. Qi, G. Pontius, and J. Cornell. 1995. Modeling spatial and temporal patterns of tropical land-use change. *Journal of Biogeography* 22:753–757.

Kirtland D., L. DeCola, L. Gaydos, W. Acevedo, K. Clarke, and C. Bell. 1994. An analysis of human induced land transformations in the San Francisco Bay/Sacramento area. *World Review* 6 (2):206–217.

White, R., and G. Engelen. 1994. Cellular dynamics and GIS: Modeling spatial complexity. *Geographical Syste*ms 1:237–253.

———. 1993. Cellular automata and fractal urban form: A cellular modeling approach to the evolution of urban land-use patterns. *Environment and Planning* 25:1175–1199.

Section III: Visualization, Representation, and Communication

Texture as a Property of Remote-sensed Images: Augmenting Standard Spectral Classification Techniques Identification of Built Patches on the Upper San Pedro Basin Landscape

Ward Brady and Ryan Miller

Introduction

Classification of remote-sensed images of semiarid and arid landscapes has proven a difficult and sometimes not very accurate task. At the same time, the need for accurate and up-to-date maps of man-made and natural features is greater than ever as small settlements explode into cities, consume resources, and cover land with cement and asphalt over a few short years. This has particularly been the case with the Upper San Pedro Basin (USPB) when attempting to evaluate indicators of watershed health obtained from remote-sensed imagery. Anthropogenic change on the landscape, for the purpose of indicator development in the USPB, has been divided into the built environment (urban, suburban, rural, and industrial development along with the necessary infrastructure) and agriculture. Both these landscape features are critical for development of indicators that are ecologically meaningful as well as statistically reliable. Of the two landscape features, the built environment has proven to be the more difficult to delineate.

Many studies involving the use of textural information alone were completed to classify images, some of which included the use of spectral information (Haralick et al. 1973; Hsiao and Sawchuck 1989). However, only a handful have used the two together to classify an image, and fewer still have done this for the purposes of classifying land use in a semiarid environment (Abeyta and Franklin 1998; Peddle and Franklin 1991; Ryherd and Woodcock 1996).

The objectives of this paper are first to illustrate several different approaches for describing texture of remote-sensed imagery, and second to illustrate how textural information, added to the spectral information, can increase the accuracy of land-use classification.

The Upper San Pedro Basin

The Nature Conservancy, in cooperation with over 100 private and public partners, has declared the Upper San Pedro Basin to be one of 12 last great places of the western hemisphere in terms of ecological diversity and importance (Nature Conservancy 2001). One rationale for this designation is the very high faunal diversity of the watershed (for instance, over 300 bird species occur on the San Pedro riparian corridor alone). In addition, the San Pedro is among the few western rivers with an unregulated flow. The San Pedro Riparian National Conservation Area is a 23,000-hectare natural area established in 1988. The Nature Conservancy also operates an internationally known bird sanctuary, the Ramsey Canyon Preserve, in the Huachuca Mountains and the National Audubon Society manages a unique research station, the Appleton-Whittell Research Ranch Sanctuary, on the watershed's grasslands.

The San Pedro watershed is a semiarid region located approximately 110° west longitude and 31° north latitude (Figure 1). The river flows north into the United States from its headwaters near Cananea, Sonora, Mexico. The upper basin has a total area of about 6,500 square kilometers with about 72 percent of the area occurring in the United States. Elevations range from 1,100 to 2,900 meters while annual rainfall varies from about 300 millimeters to 750 millimeters. Within this relatively small watershed an extraordinary diversity of ecosystems exists including Chihuahuan Desertscrub, Semidesert Grassland, Riparian Deciduous Forest, Chaparral, Madrean Evergreen Forest and Woodland, and Madrean Montane Conifer Forest.

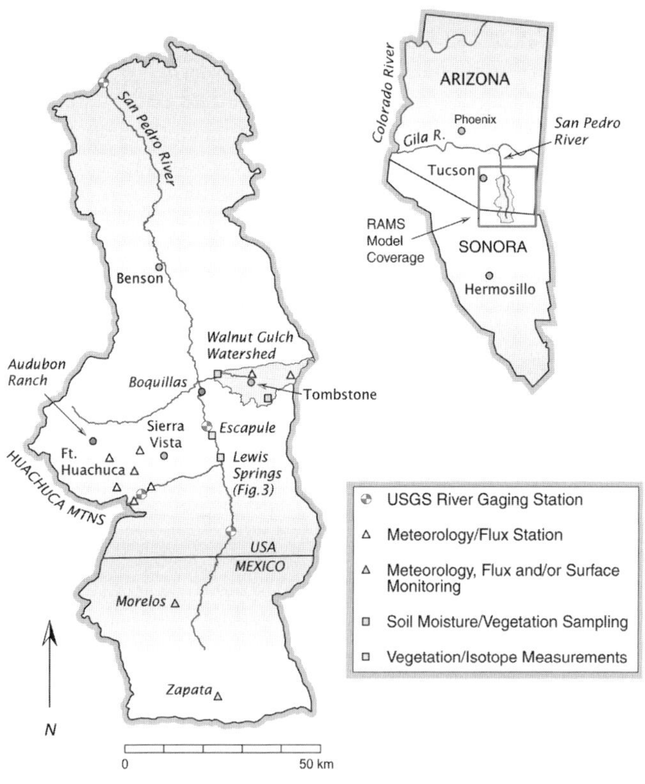

Figure 1. Location of the Upper San Pedro Basin.

Remote-sensed imagery used in this study was taken from the Landsat 5 satellite during September 1992. The scene was selected because of the existence of orthophotos from the same year, which were used for ground-truthing. Seven spectral bands were originally obtained. Band six was discarded because of its low spatial resolution. For purposes of this paper, a small image located in the area of Sierra Vista, Arizona, (Figure 2) was used. The image had a northwest coordinate (UTM) of (579234.0, 3493901.0) and a southeast coordinate of (564642.0, 3479309.0). The area of the image was 21,292.7 ha.

Texture

It is easy to agree that an object or scene has texture. Not so often agreed upon is the actual definition of texture. It is even more difficult is to quantify texture with the use of computers. Texture may be defined as the spatial pattern of gray levels in an image (Ryherd and Woodcock 1996). The characteristics that an image or object must possess in order to have texture can be summed up two different ways: for an object in an image to have texture, it must posses contrasting levels of gray with both directionality and regularity across the surface, and texture in an image has the presence of completely random gray levels (Haralick et al. 1973). The latter definition differs from the first in that a texture may be completely chaotic in nature, as opposed to order or regu larity present in the first definition. For example, a woven piece of cloth has texture. This texture is a product of under and overlapping threads which, when exposed to a source of light in any specific angle and azimuth, will posses regularly varying gray levels in any one direction. In contrast, a pile of sand possesses a countless number of particles in completely random order. This pile also has texture and the varying levels of gray are revealed in the same fashion as the piece of cloth. However, unlike the piece of cloth, the texture of the pile of sand is chaotic in nature.

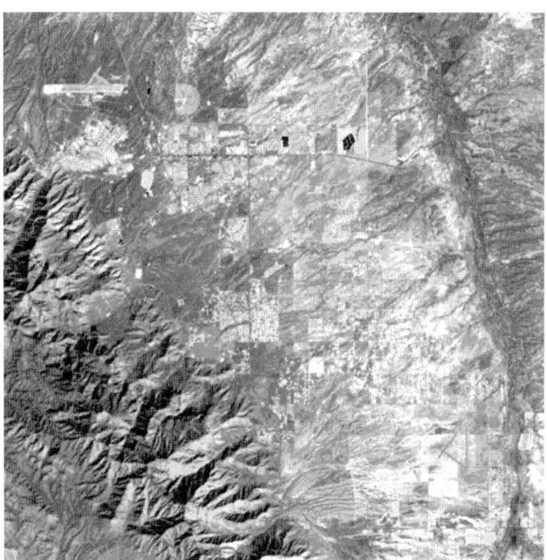

Figure 2. Composite of the Sierra Vista region of the Upper San Pedro Basin.

Describing texture has been much easier than quantifying it (Haralick et al. 1973). In fact, texture has been described in many ways including rough, fine, smooth, velvety, coarse, and bumpy. A number of studies quantify texture, most for the purpose of segmenting or classifying an image (Haralick et al. 1973; Mitchell et al. 1996; Hsiao and Sawchuck 1989). The approaches and algorithms used to accomplish these tasks are almost as numerous as the adjectives used to describe texture itself. Some examples include Fourier transformations, occurrence and co-occurrence matrices, fractals, mathematical morphology (Gose et al. 1996) and, recently, wavelets.

For example, the following figures illustrate the use of several algorithms to describe the texture of the study area shown in Figure 2. Note that the built portions of the landscape seem to have textural properties that distinguish them from the surrounding landscape.

A fragmentation algorithm (Figure 3) clearly illustrates that the built portions of the landscape tend to have higher fragmentation values (brighter colors) than the surrounding natural vegetation.

Figure 3. Description of the study area using a fragmentation algorithm.

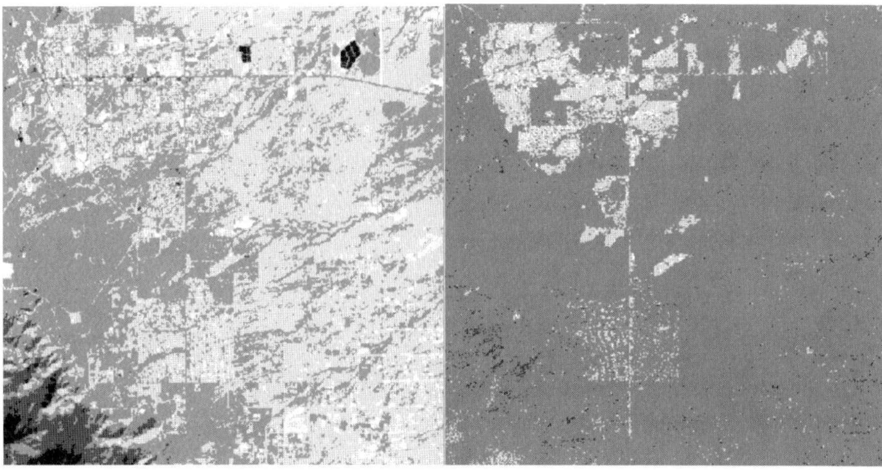

Figure 4. First and second principal components of the study area image. (caption for combined figures 4a and 4b).

The first and second principal components (Figure 4) of the study area image also suggest that the built portions of the landscape have image properties that are very different from the surrounding landscape. In this case, these properties are a combination of spectral reflection and texture. Note how built portions of the landscape are highlighted in the second component.

When looking at the landscape on a slightly larger scale, some features of the built landscape have spectral properties that are very similar to features in the natural landscape. This is particularly the case in semiarid and arid landscapes where significant portions of the ground surface (even in undisturbed condition) are bare soil. For the San Pedro example, areas near the San Pedro river (having significant exposed soil) are very difficult to distinguish from much of the built area in Sierra Vista (Figure 5).

Figure 5. Classification of an image surrounding Sierra Vista showing forest on the Huachuca Mountains (light area to left), the San Pedro River (medium gray curvilinear shape on image right), croplands (light gray to image right) and the built landscape/rock and bare soil complex (medium gray).

Figure 6. Image segmentation using spatial signature of the river terraces. The darkest gray pixels (see upper right of image) indicate higher probabilities of match with the river terrace spatial signature.

However, even if these areas share spectral properties, they differed in the spatial arrangement. Therefore, one approach to distinguishing these features is the use of spatial signatures. A spatial signature is a description of the spectral reflectance of a group of pixels which typify the feature of interest. The spatial signature of the feature can then be compared to the spatial signature of tiles which are moved across the image of interest and the similarity of signatures can be recorded in a new image. Figure 6 illustrates an image segmentation using spatial signatures of the river terraces where signatures were compared using a standard Euclidian distance algorithm. This image then assists in distinguishing between the built landscape/rock and bare soil complex illustrated in Figure 5.

Studies done in the last few years have shown that including texture in the classification process has its rewards. Ryherd and Woodcock combined spectral and textural data to segment three discernibly different images, including one simulated image of a coniferous forest, a natural vegetation area image, and an image of a mixed-use suburban area. In their study, they used a multi-pass, pair-wise, region-growing algorithm to segment the images and found that the addition of texture to segmentation "can improve the accuracy in areas where the features of interest exhibit differences in local variance (1996)."

Peddle and Franklin (1991) found that by using a co-occurrence matrix and measuring the angular second moment, entropy, and inverse difference moment, along with other ancillary data, they could increase the classification accuracy of nine land use classes. Thcy accomplished this by grouping several combinations of spectral data from either SPOT or SAR and the texture information and ancillary data together. The groups were then assessed for accuracy via stepwise and linear discriminate analysis procedures.

Bruzzone et al. (1997) also found increased classification accuracies from the addition of information derived from a co-occurrence matrix. They applied the matrix on band 5 of Landsat images of rural areas of central Italy and had an increase in the Kappa Coefficient from 0.529 for the spectral data alone to 0.616 for the spectral data and texture data. Classification of their data was accomplished through the use of a modified maximum-likelihood approach.

Land-use class selection

Eight land-use classes were selected to represent the major landscape features within the study area (Table 1). The high density urban class included most of Sierra Vista proper and some surrounding communities and consisted of tightly packed residential and commercial areas whose individual areas may be small and have little to no vegetation present or relatively large, as in the case of a strip mall, also with little to no vegetation cover. In both cases, the dominant and visible materials are man-made and include such things as pavement, cement, tile, and metal. The low-density urban class encompassed outlying communities around Sierra Vista, which may include planned communities and smaller ranchettes located closely together. Some distinguishing characteristics of this land use class include larger individual plots with the natural vegetation often removed mechanically or with livestock; the presence of introduced vegetation such as lawn grass, which is kept alive throughout the year and man-made materials like those in the high-density class, but with much less overall ground coverage. Large shopping and industrial areas are not present in these areas.

The built landscape is defined (in our study) as the union of the high- and low-density urban areas. The term *urban* without any modifier is synonymous with the term *built landscape*. High vegetation cover includes much of the Huachuca military reservation and is identified by desert scrub and an abundance of grass. These areas exist primarily due to the exclusionary characteristic of the reservation, which does not allow the grazing of domestic livestock within its boundaries. The low vegetation cover class, on the other hand, includes areas east and southeast of Sierra Vista that have less grass cover due to the lower elevation and grazing. Much of this land has been divided up into ranchettes, often with domestic animals present. The water class exists primarily as a large waste treatment plant east of Sierra Vista. The agriculture class is primarily composed of wheat and alfalfa. Both the shade class and forest class are located in the upper elevations of the Huachuca Mountains.

Table 1. Eight feature classes used as training areas for supervised classification.

Land use	Description
High-density urban	Compact residential and commercial
Low-density urban	Multi-hectare residential plots and small ranchettes
High vegetation cover	Abundant grasses and desert scrub
Low vegetation cover	Bare ground with sparse vegetation
Water	Wastewater treatment plants
Agriculture	Wheat and alfalfa fields
Shade	Shaded slopes
Forest	Oak and pine

Figure 7. Grayscale image of a color composite of bands 1, 4, and 5 used in entropy measurements.

Figure 8. A texture image based on entropy calculation from a gray level occurrence matrix (tile size 17 X 17).

Texture classification

The three Landsat bands with the lowest overall correlation (1, 4, and 5) were combined to produce a color composite, which was subsequently converted into an eight-bit binary gray scale image. This grayscale image was then used in the analysis of texture (Figure 7).

Image classification

Results of the image processing provided a series of images representing textural information and classified images based on texture and spectral information. Figure 8 is an example of an entropy image (with a distance of 4, tile size of 17 x 17, and angle of 0).

Figure 9 shows the result of image segmentation and reclassification of the original eight feature classes into two classes: urban and non-urban. Accuracy assessment of this image indicates that the area identified as urban (built) included many natural landscape features that are spectrally similar to the actual built landscape. The overall accuracy of the image was 63 percent.

Figure 10 shows the result of image segmentation when textural information (entropy image illustrated in Figure 4) was included along with spectral information. Again, the original eight feature classes were combined into urban (the built landscape) and other. The result was a significant ($p = 0.05$) increase in image accuracy. The principal change was the elimination of non-built landscape features resulting in an overall accuracy of 87 percent.

Figure 9. Urban and non-urban classes from spectral alone classification technique.

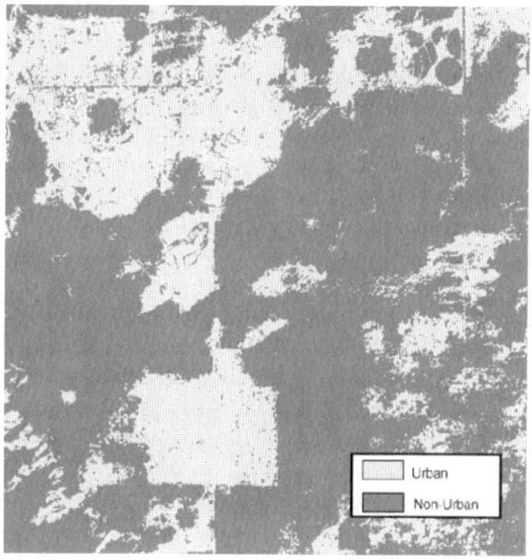

Figure 10. Urban and non-urban classes from 17l4 classification technique.

Table 2 shows the results from the accuracy assessment of the classification techniques compared to known urban points. The contingency table test resulted in rejection of the null hypothesis, and a conclusion that the number of built landscape points correctly classified differed between classification techniques (p = 0.05, X229 = 4184.00).

Dunnet's test of proportions indicated that several of the techniques were significantly different from the spectral alone in their ability to classify urban land use in any of the data sets (p = 0.05, Table 3). In most cases, the larger-sized tiles improved the classification accuracy the greatest, which may be a function of the pattern distribution of the urban areas. Additional tests would have to be performed to determine if this is indeed the case. Therefore, the results of this study indicate that the use of textural information derived from the calculation of entropy did increase the accuracy of classification of urban areas in several of the cases.

It should be noted that the 20 cases where the accuracy increased are from the larger tile sizes and that the technique to produce the best results was of the largest tile size. In future studies of this type it may be seen that increasing the tile size even further will produce even better results.

As for those techniques that were found not to be significantly different than spectral alone numerous reasons may exist, only a few will be discussed here. First, another method of obtaining an appropriate grayscale image may be in order. The method used in this study was loosely founded on the correlation of three of the spectral bands to the rest. Nothing states that three image bands are necessary or even two. For example, the first band of a principal components analysis may be used, or a single band from the spectral data as in the study conducted by Bruzzone et al. (1997), who used only a single Landsat spectral band to quantify texture. The options here are also endless, and because this field is relatively new, no standard has yet been set.

Table 2. Proportion classified urban or non-urban for points known to be urban and each technique. Total number of points: 208. Null hypothesis: classification proportions are the same for each technique (Ho: P3|1 = P3|2 = … = P17|4 = PS1A).

Technique	**Spectral Alone**	**3|1**	**3|2**	**5|1**	**5|2**	**5|3**	**7|1**	**7|2**	**7|3**	**7|4**	**9|1**	**9|2**	**9|3**	**9|4**	**11|1**
Urban	0.63	0.68	0.66	0.73	0.70	0.74	0.78	0.86	0.75	0.79	0.79	0.81	0.82	0.80	0.79
Non-Urban	0.37	0.32	0.34	0.27	0.30	0.26	0.22	0.14	0.25	0.21	0.21	0.19	0.18	0.20	0.21
Technique	**11|2**	**11|3**	**11|4**	**13|1**	**13|2**	**13|3**	**13|4**	**15|1**	**15|2**	**15|3**	**15|4**	**17|1**	**17|2**	**17|3**	**17|4**
Urban	0.82	0.84	0.84	0.80	0.84	0.86	0.86	0.73	0.86	0.86	0.73	0.82	0.86	0.87	0.86
Non-Urban	0.18	0.16	0.16	0.20	0.16	0.14	0.14	0.27	0.14	0.14	0.27	0.18	0.14	0.13	0.14

Table 3. Critical values (q) calculated from the Dunnet's test for those techniques found to be significantly different from the control. Null hypothesis: Each classification proportion is the same as the proportion for the spectral alone technique (Ho: P3|1 = PS1A, P3|2 = PS1A, … = P17|4 = PS1A). q0.05, infinity, 29 = 3.448

Technique	**Spectral Alone**	**7|2**	**7|4**	**9|1**	**9|2**	**9|3**	**9|4**	**11|1**	**11|2**	**11|3**	**11|4**
q =	Control	5.505	3.576	3.696	4.062	4.311	3.939	3.576	4.311	4.826	4.826
Technique	**13|1**	**13|2**	**13|3**	**13|4**	**15|2**	**15|3**	**17|1**	**17|2**	**17|3**	**17|4**	
q =	3.939	4.959	5.505	5.505	5.505	5.366	4.438	5.505	5.647	5.366	

Second, entropy may be the wrong way to quantify texture in the first place. A number of options are available, including other co-occurrence matrices, occurrence matrices, fractals, and mathematical morphology, to name but a few. Ryherd and Woodcock (1996) implemented a multi-pass, pair-wise, region-growing algorithm to successfully segment three discernibly different images. Recent advancements in signal processing and data compression such as wavelets may also be a promising addition to the standard spectral analysis. By being able to decompose an image to a point, similar features may be identifiable from one image to another. For example, the layout of two neighborhoods may, at first glance, be different, but at some point in decomposition of the image of each of them, they may be very similar mathematically as cell values become more generalized and the overall image becomes more homogeneous.

Lastly, it is likely that the texture band is not the source of confusion, but rather the spectral bands. Often, what the human eye might consider an urban feature would, when examined by a computer, be classified as some non-urban feature (Campbell 1990). The reason for this is that the makeup of urban landscape is heterogeneous in nature. Take, for example, a residential neighborhood. If it were entirely made of concrete (i.e., concrete roads, houses, etc.), separating it from the surrounding environment would be a simple task, but since neighborhoods comprise such materials as wood, dirt, asphalt, and concrete, to name a few, the task becomes much more complicated. When exposed to an image of a cityscape, the human eye quickly delineates such features as residential neighborhoods. It does this through a number of processes and by experience. These processes include comparing the frequency and distribution of light and dark features of an area and separating them from those that differ. The nature of standard spectral classification techniques like the one used in this study is that the spectral signature of each pixel in the scene is determined and compared to the training signature produced by the user. When classifying these pixels these techniques do not take into account the surrounding pixels and so, if a match is made between a pixel within an urban area and a signature developed for land use outside urban areas, an error has been produced only as far as the user may disagree with what the technique considers urban land use.

Conclusion and Recommendations

In this study, the use of textural information based on entropy to improve the classification of urban land features in a semiarid environment was shown to be effective. In 20 variations of the entropy calculation, a significantly more accurate classification than that of using spectral bands alone was achieved. The spectral technique was able to correctly classify 63.0 percent of the known urban points while the 20 techniques found to be significant had accuracies ranging from 72.6 percent to 86.5 percent.

The results of this study are in line with other similar studies. Ryherd and Woodcock (1996) found the addition of spatial attributes, like texture, to the segmentation of their images increased the overall accuracy. Gong and Howarth (1990) included image structural data to their classification of the northeastern fringes of Toronto, Canada, and had an increase in classification accuracy 10 percentage points over standard spectral classification techniques. Peddle and Franklin (1991) were able to increase their classification accuracy of nine land cover classes by adding textural information to their classification technique.

Although this study by no means produced perfect results, it does provide a foundation from which to explore the use of texture as a standard component in the classification of remote-sensed data.

References

Abeyta, Andres M., and Janet Franklin 1998. The accuracy of vegetation stand boundaries derived from image segmentation in a desert environment. *Photogrammetric Engineering and Remote Sensing* 64 (1):59–66.

Bruzzone, L., C. Conese, F. Maselli, and F. Roli. 1997. Multisource classification of complex rural areas by statistical and neural-network approaches. *Photogrammetric Engineering and Remote Sensing* 63 (5):523–533.

Campbell, J. B. 1990. *Introduction to remote sensing*. 2nd ed. New York: The Guilford Press.

Gong, Peng, and Philip J. Howarth. 1990. The use of spectral information for improving land-cover classification accuracies at the rural-urban fringe. *Photogrammetric Engineering and Remote Sensing* 56 (1):67–73.

Gose, Earl, Richard Johnsonbaugh, and Steve Jost. 1996. *Pattern recognition and image analysis*. Upper Saddle River, N.J.: Prentice Hall, Inc.

Haralick, Robert M., K. Shanmugam, and Its'hak Dinstein. 1973. Textural features for image classification. *IEEE Transactions on Systems, Man, and Cybernetics* 3 (6):610–621.

Hsiao, John Y., and Alexander A. Sawchuk. 1989. Unsupervised textured image segmentation using feature smoothing and probabilistic relaxation techniques. *Computer Vision, Graphics, and Image Processing* 48:1–21.

Mitchell, O. R., C. R. Meyers, and W. Boyne. 1996. A max-min measure for image texture analysis. *IEEE Transactions on Computers* C-25 (4):408–414.

Nature Conservancy. 2001. Web document available at <http://www.tncarizona.org>. Viewed July 22, 2001.

Peddle, Derek R., and Steven E. Franklin. 1991. Image texture processing and data integration for surface pattern discrimination. *Photogrammetric Engineering and Remote Sensing* 57 (4):413–420.

Ryherd, Soren, and Curtis Woodcock. 1996. Combining spectral and texture data in the segmentation of remote-sensed images. *Photogrammetric Engineering and Remote Sensing* 62 (2):181–194.

Reeves, Mathew. 1998. Mapping Sonoran desert vegetation of the McDowell Mountains using hyperspectral imagery. Master's thesis, Arizona State University.

Zar, Jerrold H. 1996. *Biostatistical analysis*. 3rd ed. Upper Saddle River, N.J.: Prentice Hall, Inc.

Balancing Measurement Precision with Cognitive Efforts in Weighting Method Selection

Wei-Ning Xiang

Major Weighting Methods

Typically, a weighting method comprises two parts of elicitation and representation. It first elicits users' perceptions about the importance or relative importance of map layers, then expresses them in mathematical forms. Accordingly, various weighting methods can be grouped into three broad categories on the way they elicit users' perceptions: direct assessment, tradeoff weighting, and equal weighting.

The direct assessment approach requires a person to quantitatively state either the importance or relative importance of each map. Methods under this approach include, but are definitely not limited to, ranking, rating, ratio questioning, and Metfessel allocation. In ranking, a person ranks all the maps according to their relative importance. The higher a map's rank position, the more important it is. The actual importance (weight) value of each map is then determined mathematically on the map's rank position. In rating, a person rates each map's importance on a scale of 0 to 1 (or equivalently, 0 to 10, 0 to 100), with 1 being the highest level of importance. In ratio questioning, a person compares two maps at a time and answers such question as "how much more important is one map over the other, with 1 being equally important, 2 slightly more important, 3 moderately more important . . .?" A transformation function is then executed to calculate maps' weights based on the answers. The Metfessel allocation method asks one to allocate 100 points among the maps in proportion to their importance. Despite their popularity and, especially for rating and Metfessel allocation, ease of use, methods under this approach suffer from the ambiguity in the definition of importance, and thus cannot guarantee the theoretical validity of the derived weights (Hobbs 1980, 727–728; Rigley and Rijsberman 1994, 26).

The tradeoff weighting approach does not require a person to assign weights to, nor state relative importance of, the attributes or criteria (in the case of map overlays, maps) directly. Instead, it asks one to state how much compromise he is willing to make between two attributes or criteria when an ideal combination of the two is not attainable. A typical question, as Keeney and Raiffa say, is "how much achievement on objective 1 is the decision maker willing to give up in order to improve achievement on objective 2 by some fixed amount?" (1976, 66). This approach guarantees the theoretical validity of the derived weights. However, its practical applicability has been modest because of the demanding requirement that a person must be able to make explicit and consistent tradeoffs (Hobbs 1980, 729; Lai and Hopkins 1989, 167; Rigley and Rijsberman 1994, 26). In the literature, there have been no reports on its applications to land suitability assessment or weighted map overlays.

Contrary to the first two approaches that seek differential weights, that is, weights of different numerical values, the equal weighting approach requires a person's consent that all the contributive map layers are equally important and assigns a same weight value to all the layers. Despite its obviously counterintuitive nature, some advocates argue that equal weights frequently produce assessment results that are as good in quality while sparing one the cognitive effort of estimating a precise set of differential weights (Dawes and Corrigan, 1974). However, the more recent studies have shown that the equal weighting approach is easier but also substantially inferior to the other two approaches in producing quality results (Jia et al. 1998, 102; Newman 1977, 313; Stillwell et al. 1981, 63–64, 75–76).

Measurement Precision and Cognitive Effort

The above three approaches are all designed for elicitation purposes, but at different precision levels. The equal weighting approach solicits users' judgment on a nominal scale about whether a map is contributive or not. It then assigns all the contributing map layers an equal weight. The direct assessment approach elicits perceptions either on an ordinal scale about map layers' contribution shares (in the cases of ranking and Metfessel allocation), or on an interval scale about the pairwise contributions (in the case of ratio questioning). It then transforms the perceptual information into weights on a ratio scale through a mathematical standardization. The rating methods and the tradeoff weighting approach operate both on a ratio scale but the latter requires more explicit and usually consistent input from users.

Methods of different precision levels have different requirements for cognitive effort and time commitment from the users. By *users* it is meant two groups of people: those who use a weighting method to solicit others' perceptions and those who express their own preferences through a weighting method. For both groups, generally speaking, the more precise the method, the greater the effort it requires and the more time-consuming it is. The tradeoff approach is the most effortful and time-consuming, equal weighting the most effortless, and the direct assessment approach lies in between (Jia et al 1998, 102). For instance, in a policy-making process that involved 11 attributes, it took eight hours for Keeney to assess the utility function of a single person by the tradeoff approach (Newman 1977, 313). In a land suitability assessment project that involved 24 parent map layers it took Whitley and Xiang an average of four hours to solicit weights through a direct assessment approach from each of the five human experts (Xiang and Whitley 1994, 290). In both cases, the amounts of time and effort spent on questionnaire preparation, human expert identification, initial contacts, follow-ups, and travel are not counted.

Measurement Precision and Insensitivity

Is it worthwhile to spend more time and effort on the pursuit of more explicit perceptual information for weighting purposes? It depends on one's expectations for the precision levels in the overlay results. The answer is yes when the major concern is about the cardinal differences among the sites in their suitability assessment scores. As discussed above, weights are highly dependent on the elicitation method, their numerical values vary substantially from one method to another even with the perceptual information from a same person. Once incorporated into the land suitability assessment process, every different weight value will produce an overlay map with distinctive results on a cardinal scale. Detailed numerical values will delineate specific differences among the sites, and therefore are worth the effort. However, if the main consideration is given to the ordinal differences, the answer becomes maybe or even no because of the ordinal insensitivity of map overlays towards weighting. This is a phenomenon when a site suitability ranking on an overlay map remains unchanged for substantial variations in the weights.[1] To illustrate, consider the overlay maps in Figure 1 (Xiang and Salmon 2001). Each of the nine overlay maps is a weighted combination of two parent maps, 1 and 2. When map 1 is assigned a weight value of 0.1, that is, W1 = 0.1, map 2 automatically receives a weight value of 0.9, that is, W2 = 0.9. Once incorporated into a map overlay process, these two values produce an overlaid map with a site ranking S11 > S21 > S31 > S12 = S14 > S22 = S24 > S32 = S34 > S13 > S23 > S33 (the first overlay map in Figure 1). The ranking is not changed for the variations in W1 values from 0.1 to 0.2 even to 0.3 (0.9, 0.8, and 0.7, respectively, for the W2 values). Symmetrically, also in Figure 1, a different site ranking S11 > S12 = S14 > S13 > S21 > S22 = S24 > S23 > S31 > S32 = S34 > S33 is found on three overlay maps that correspond to W1 values of 0.7, 0.8, and 0.9, respectively. This phenomenon can also be described as *weight segmentation* (Xiang 2000a; Xiang and Salmon 2001). That is, the weight value range, usually defined on [0, 1], is actually segmented. Each segment comprises weight value(s) that, once incorporated into the map overlay process, will produce an ordinal set of suitability scores under a unique order of dominance. In the above example, the three W1 values 0.1, 0.2, and 0.3, yield three overlay maps of the same ordinal result because of their positions in the same segment. So do the W1 values 0.7,

[1] Edwards (1977) first observed this phenomenon in decision analysis. Xiang (2000a, 607–611) discusses in great detail the phenomenon in land suitability assessment which he referred to as *weight segmentation*. Xiang and Salmon (2001) proposed a scheme of computer-button design for map overlays based on this phenomenon.

Figure 1. The ordinal insensitivity of map overlays towards weights

W1	Weighted map overlay and suitability score calculation			Site rankings based on the suitability scores
0.1	1.00	0.95	0.90	
	0.55	0.50	0.45	S11>S21>S31>S12=S14>S22=
	0.10	0.05	0.00	S24>S32=S34>S13>S23>S33
	0.55	0.50	0.45	
0.2	1.00	0.90	0.80	
	0.60	0.50	0.40	S11>S21>S31>S12=S14>S22=
	0.20	0.10	0.00	S24>S32=S34>S13>S23>S33
	0.60	0.50	0.40	
0.3	1.00	0.85	0.70	
	0.65	0.50	0.35	S11>S21>S31>S12=S14>S22=
	0.30	0.15	0.00	S24>S32=S34>S13>S23>S33
	0.65	0.50	0.35	
0.4	1.00	0.80	0.60	
	0.70	0.50	0.30	S11>S21>S12=S14>S31>S22=
	0.40	0.20	0.00	S24>S13>S32=S34>S23>S33
	0.70	0.50	0.30	
0.5	1.00	0.75	0.50	
	0.75	0.50	0.25	S11>S12=S14=S21>S13=S22=
	0.50	0.25	0.00	S24=S31>S32=S23=S34>S33
	0.75	0.50	0.25	
0.6	1.00	0.70	0.40	
	0.80	0.50	0.20	S11>S12=S14>S21>S13>S22=
	0.60	0.30	0.00	S24>S31>S23>S32=S34>S33
	0.80	0.50	0.20	
0.7	1.00	0.65	0.30	
	0.85	0.50	0.15	S11>S12=S14>S13>S21>S22=
	0.70	0.35	0.00	S24>S23>S31>S32=S34>S33
	0.85	0.50	0.15	
0.8	1.00	0.60	0.20	
	0.90	0.50	0.10	S11>S12=S14>S13>S21>S22=
	0.80	0.40	0.00	S24>S23>S31>S32=S34>S33
	0.90	0.50	0.10	
0.9	1.00	0.55	0.10	
	0.95	0.50	0.05	S11>S12=S14>S13>S21>S22=
	0.90	0.45	0.00	S24>S23>S31>S32=S34>S33
	0.95	0.50	0.05	

0.8, and 0.9. Now if the value 0.1 is derived from an easier method and 0.3 from a harder one, is it worthwhile to use the latter method over the former? Of course not.

Zipf's Principle of Least Effort

Therefore, if ordinal differences in the overlay results are the major concern, the extra effort and time devoted to a more precise weighting exercise become worthless when the weight values do not cause any discrepancy in site ranking from those by the less precise weight values. But then the question is: will people be willing to give up the pursuit for higher precision and use a less rigorous weighting method? More than likely, they will. According to Zipf (1949, 5–8), a basic principle that governs all varying conduct of an individual is that of least effort. Stated informally, all else being equal, people tend to use as little effort as necessary in reaching a desired result (Jia et al. 1998, 102; Payne et al. 1993, 13; Zipf 1949, 5–8). In our case, therefore, there is no reason why people are not willing to use a less effortful and time-consuming method if they are convinced that it will produce a desirable and/or essentially same result.

Support for a Least Effort Selection: A Prototype

If we plan to build a computer-based information system to help people select the weighting method of least effort, what would the system look like? First, it should provide a spectrum of weighting methods. This is a major point of departure from most map overlay support systems, for they are usually committed to only one method. Second, it should provide information about the different methods' precision levels and their effort and time requirements. Third, it should have information about the ordinal insensitivity of map overlays toward weights derived from different methods.

Shown in the appendix to this article is a prototype of such a system. It includes four methods representing the three elicitation approaches, and organizes them in a precision/effort space. The equal weighting method (Dawes and Corrigan 1974) represents the equal weighting approach; the rank-order centroid method (Barron and Barrett 1996) and AHP (Saaty 1980) represent the direct assessment approach; and the tradeoff method (Keeney and Raiffa 1976; Xiang 2000b) represents the tradeoff weighting approach. In the prototype, the user can select a weighting method in two ways. Under the first option, he can choose the method he/she likes to use given the knowledge about its position in the precision/effort space. Under the second option, he will be guided by heuristics to navigate from the easiest equal weighting toward the toughest tradeoff weighting and stop at a point where he/she feels satisfied with the overlay results. The system is currently under construction and will incorporate the ordinal insensitivity information for each method.

Conclusion

In the design and development of computer-based support systems for weighting purposes, an important yet often neglected aspect is the tradeoffs people make in practice between measurement precision and cognitive effort. In selecting a weighting method, people tend to follow Zipf's Principle of Least Effort and balance, either systematically or intuitively, precision level of a method with the amount of effort and time it requires. A system design that provides information about the tradeoffs will help people select a least-effort method. The prototype presented in this article sets up the stage for further investigation.

Acknowledgements

Some of the ideas in this article were first presented at The Symposium on Integrated Land Use and Environmental Models: A Survey of Current Applications and Research, October 20–21, 2000, Arizona State University, Tempe, Arizona. I wish to thank Subhro Guhathakurta at Arizona State University for inviting me to the symposium, and Lew Hopkins at University of Illinois for his comments on my presentation at the symposium. I am also indebted to Bryan Townsend at University of North Carolina at Charlotte for programming the prototype and preparing the appendix.

References

Barron, F. H., and B. E. Barrett. 1996. Decision quality using ranked attribute weights. *Management Science* 42:1515–1525.

Dawes, R. M., and B. Corrigan. 1974. Linear models in decision making. *Psychological Bulletin* 81:95–106.

Edwards, W. 1977. How to use multiattribute utility measurement for social decisionmaking. *IEEE Transactions on Systems, Man, and Cybernetics* 7(5):326–340.

Hobbs, B. F. 1980. A comparison of weighting methods in power plant siting. *Decision Sciences* 11:725–737.

Jia, J., G. W. Fischer, and J. S. Dyer. 1998. Attribute weighting methods and decision quality in the presence of response error: A simulation study. *Journal of Behavioral Decision Making* 11:85–105.

Keeney, R. L., and H. Raiffa. 1976. *Decisions with multiple objectives: Preferences and value tradeoffs*. New York: John Wiley & Sons.

Lai, S.-K., and L. D. Hopkins. 1989. The meanings of trade-off in multiattribute evaluation methods: A comparison. *Environment and Planning B: Planning and Design* 16:155–170.

Newman, J. R. 1977. Differential weighting in multiattribute utility measurement: When it should not and when it does make a difference. *Organizational Behavior and Human Decision Processes* 54:456–476.

Payne, J. W., J. R. Bettman, and E. J. Johnson. 1993. *The adaptive decision maker*. Cambridge, U.K.: The Cambridge University Press.

Rigley, M. A., and F. R. Rijsberman. 1994. Multicriterion analysis and the evaluation of restoration policies for a Rhine estuary. *Socio-Economic Planning Sciences* 28 (1):19–32.

Saaty, T. L. 1980. *The analytic hierarchy process*. New York: McGraw-Hill.

Stillwell, W. G., D. A. Seaver, and W. Edwards. 1981. A comparison of weight approximation techniques in multi-attribute utility decision making. *Organizational Behavior and Human Performance* 28:62–77.

Xiang, W.–N. 2000a. A theoretical framework for weight value set construction in land suitability assessment. *Environment and Planning B: Planning and Design* 27 (4):599–614.

———. 2000b. A weight elicitation method for map overlays. Paper presented at The Symposium on Integrated Land Use and Environmental Models: A Survey of Current Applications and Research, October 20–21, 2000, Arizona State University, Tempe, Arizona.

Xiang, W.-N., and F. W. Salmon. 2001. Button design for map overlays. *Environment and Planning B: Planning and Design* (forthcoming).

Xiang, W.-N., and D. L. Whitley. 1994. Weighting land suitability factors by the PLUS method. *Environment and Planning B: Planning and Design* 21 (3):273–304.

Zipf, G. K. 1949. *Human behavior and the principle of least effort*. Cambridge, Mass.: Addison-Wesley.

Appendix

A. Prototype Support System for Weighting Method Selection

Figure 01. I. The user is presented with two options. He/she can select any method with the implicit knowledge about the methods' precision levels and effort requirements. He/she can also follow a heuristics to select a best suitable method. Under this second option, He/she will begin with the equal weighting method. Other methods will become available if equal weighting is not acceptable and more effort is provided to choose weights.

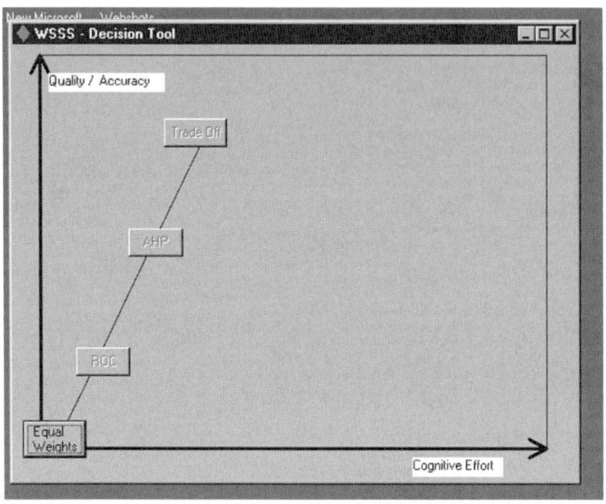

Figure 02. II. By choosing equal weights, a map is displayed.

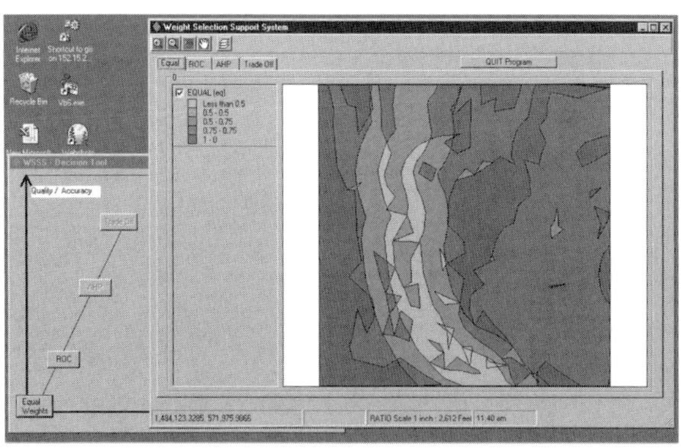

Figure 03. III. The user is presented with the option to use ROC (Rank-Order Centroid) weights if he/she is satisfied with the result.

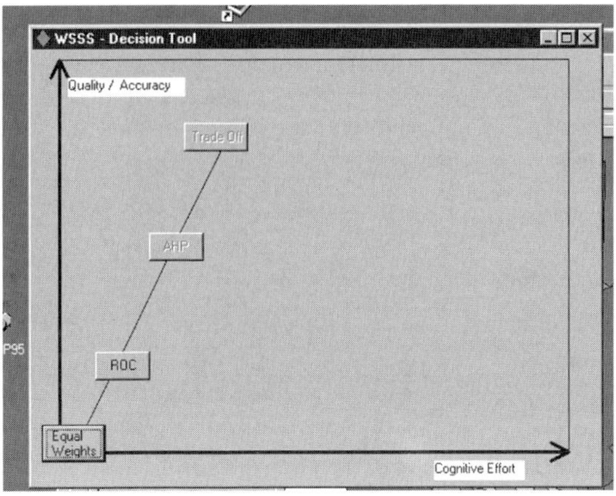

Figure 04. IV. Selection of the ROC button from the Decision Tool brings up the Elicitor Tool menu. Here the user is asked to rank the maps. The weights are immediately calculated and the user is asked to view the results if the calculated weights are satisfactory.

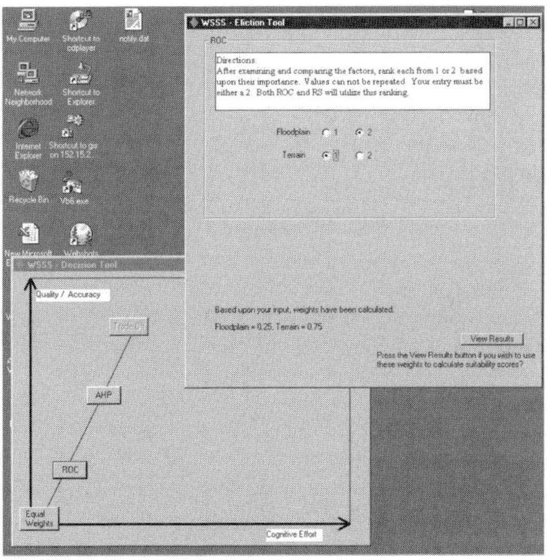

Figure 05. V. After the View results button is selected from the ROC elicitor tool menu, the resulting ROC weights are applied to the overlaid map.

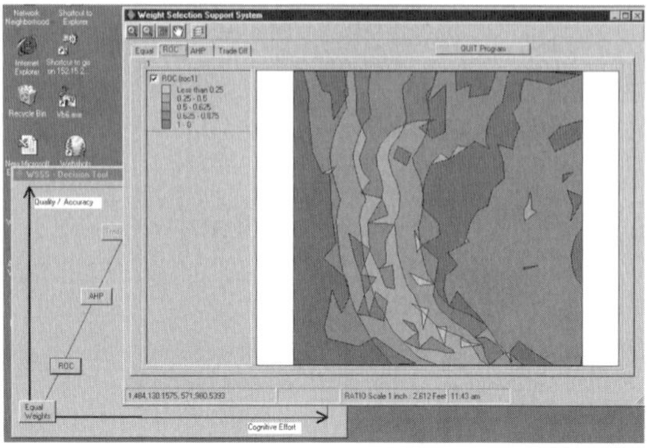

Figure 06. (no image associated with this caption). VI. If the user is still unsatisfied with the results of the ROC weights, AHP weights can be calculated.

Figure 07. VII. The WSSS Weight Elicitor tool alerts the user that they prefer one map over another and then asks the user to further clarify how much more using a slide bar and the defined scale values. The user is presented with calculated weights and asked to confirm their selection.

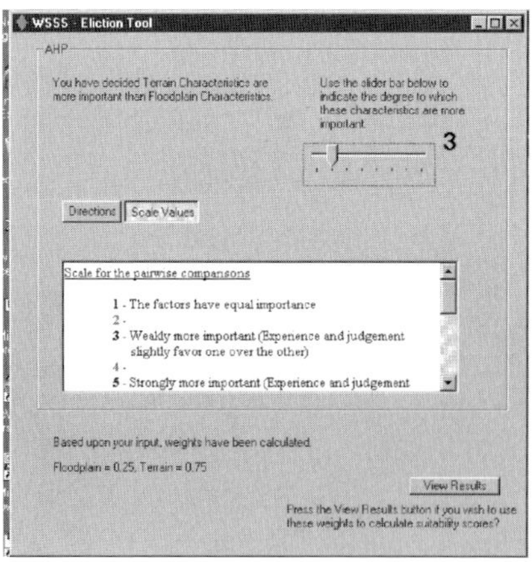

Figure 08. VIII. If the user is still unsatisfied with the results of the AHP weights, Trade Off weights can be calculated. Once the View Results button has been pressed on the AHP weight elicitor menu, the results are displayed in map form.

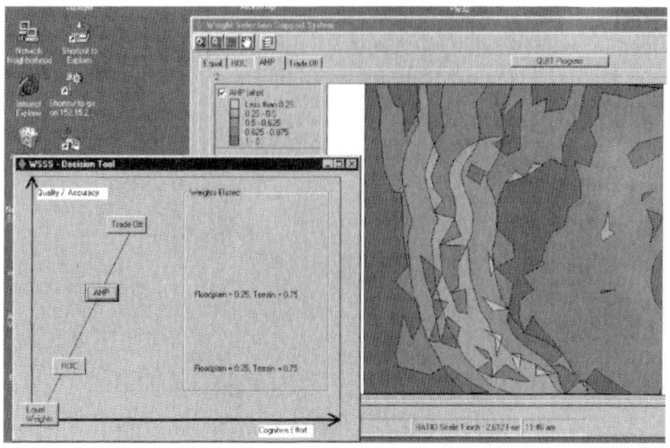

Figure 09. IX. If the user desires the optimal choice of weights the Trade Off method may be selected. The user is prompted "according to your previous choices, this is your desired weight set." The user may view other site options as well before viewing results.

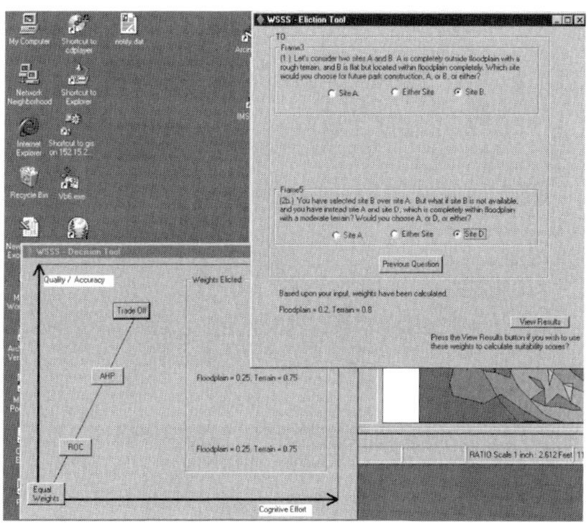

Figure 10. X. The user may view the final map depicting the results of the Trade Off weights as well as the weights elicited by previous methods.

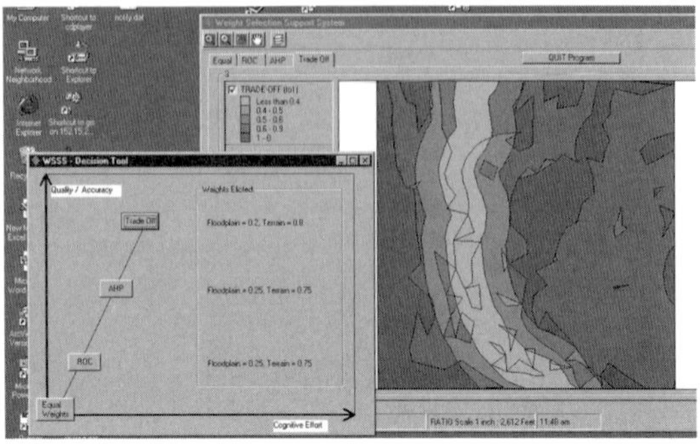

Section IV: Socioeconomic Implications of Transportation and Land Use

Modeling the Reciprocal Relationship between Metropolitan Roadway Expansion and Urban Land Development with Elementary Extensions to Environmental Consequences

Philip C. Emmi and Craig Forster

A. The Problem and its Relevance

Background

Within North American metropolitan regions, growth in urban land development, vehicular travel, and roadway capacity consistently outpace growth in either people or jobs. More puzzling is this trio's rapid growth in regions that are stable or declining in people and jobs. Disproportionate growth in urban land, roads, and traffic relative to people or jobs represents both a challenge and an opportunity.

This paper focuses on the growth dynamics of metropolitan regions. It presumes a 25-year time horizon in which to consider processes strategic to human-environmental relationships in the continent's most urbanized regions. The intention is to use a system dynamics model to articulate relationships among sectors of urban activities and to explore the policy-dependent trajectories of selected urban performance indicators.

Escalating rates of urban land development per capita have substantial impacts on the continued viability of current social, political, economic, and ecological relationships. This paper proceeds with an awareness of the need for clearer comprehension of these relationships and for opportunities to experiment with simple policy choices. It seeks to address these needs by developing a systems model of metropolitan growth dynamics that is certain of its environmental effects. It does so in a very simple and preliminary way.

Our intention is to demonstrate a modeling approach that represents the most central elements of a dynamic urban complex in a way that captures the jointly determined and simultaneously interdependent nature of key urban system components. The paper identifies and diagrams the basic patterns of urban structural interdependency. It transforms this diagram into an operational model. It then explores the response of the resulting system to selected policy initiatives.

The model presumes a prototypical urban area. Parameter values are not calibrated for any particular or specific location but are rather approximately accurate and generally representative. At this stage, we do not claim that our numerical results are empirically correct but only that the relationships among stocks and flows within the model are logically correct and properly organized to reflect the dynamics of a real-world reinforcing feedback loop. Projected trajectories of stocks and flows are not to be interpreted as prospectively credible numbers. Rather the trajectories are to be considered in relation to one another as proportionate or disproportionate to the demographic and economic drivers of the model. Thus the reader is invited to consider, for example, not only the conditions under which linear job growth get transformed into exponential growth in urban traffic volumes but also the conditions under which both jobs and traffic grow proportionately. At this stage, our intention is to lay out a system of relationships in a way that facilitates clarification, debate, and understanding.

Specifically, the model recognizes six stocks. These include basic employment, nonbasic employment, population, urban land, person trips, and roadway lane miles. For the moment, these six stocks are organized into three sectors. The first three stocks are organized into a demo-economic sector. Outputs from this sector feed into a land-use sector for computation of commercial and residential land use. Output from the land-use sector interacts reciprocally with the generation of person trips and roadway lane miles, both of which are organized into a transportation sector. Conceptual detail within each sector is kept to a minimum. Extensions to other potential sectors, such as water consumption, energy consumption, and atmospheric emissions, are readily imagined. We include an extraordinarily simplified municipal and industrial water consumption sector to illustrate patterns general throughout the urban environment. The performance of the model can be manipulated through a small set of policy variables easily altered by the user. These permit variation in demographic and economic growth, alteration in rates of roadway expansion, control of developmental densities, and expansion of mass transit capacity. The goal is to delineate the essential dynamics of an interrelated set of urban stock variables and demonstrate the possibilities for the exploration of selected policy outcomes.

A systems thinking approach

System dynamics is the technical means to engage in systems thinking. Systems thinking is not general systems theory, systems analysis, or soft systems methods. Nor is it operations research, decision analysis, or control theory. Intellectually, it has roots in objectivism, social systems theory, and policy analysis. But in its current form of practice, it actively reaches toward organizational learning and interactive decision support systems (viz., Lane 2001, Friedmann 1987). It is "the art and science of making reliable inferences about behavior by developing an increasingly deep understanding of underlying structure" (Richmond 1994, 6). "The goal of a system dynamics policy study is understanding: understanding the interactions in a complex system that are conspiring to create a problem, and understanding the structure and dynamic implications of policy changes intended to improve the system's behavior" (Richardson 1991, 162). System dynamics has its beginnings in the foundational work of Jay W. Forrester (1961, 1969, 1971).

Practitioners of systems thinking engage in a continuum of activities from the conceptual, through the analytical, to the technical. Conceptually, systems thinking involves the adoption of a perspective far enough back from its subject to allow comprehension of the enduring web of interdependencies that sustain a system's unique behavior patterns.

Analytically, systems thinking involves laying out the mechanisms that control the way the system behaves through time. This requires mapping out the four elements of system structure:

- the relevant stocks maintained by the system
- the flows that add to and reduce the stocks
- the converters that govern the flows
- the connectors that link these three other elements into an operational unit

Technically, systems thinking requires the successful manipulation of system dynamics software. We use STELLA—a Structural Thinking, Experimental Learning Laboratory with Animation, by High Performance Systems, Inc. Like other software of this genus, STELLA has a graphical user interface with four icons to represent the four elements of system structure. The user interface simplifies the development of a simultaneous set of difference equations at a level within the software where the user need not go. The software solves these equations with relatively little involvement on the part of the user. The solution produces a characteristic pattern of system behaviors or trajectories for its stock and flow values. These behaviors simulate the system's responses both to the base case and to modifications in a suite of selected policy variables.

The urban environmental problem

Three closely related flows are essential to understanding the feedback mechanism central to urban system dynamics. These are the development of urban land use (and the related but narrower process of urban property development), the generation of person and vehicle traffic, and the expansion of urban roadway capacity. These are needed to constructively articulate the linkages between urban demo-economic processes and urban environmental processes. This is so for three specific reasons.

First, an urban region's transport sector is responsible for a significant portion of its fossil fuel consumption and its generation of atmospheric emissions. Analytical treatment of these environmental and natural resource issues would be incomplete without consideration of the interrelationship between land use and transportation.

Second, within each of the nation's various climate zones, urban land developmental densities significantly affect long-term trends in municipal and industrial water demand and in the cost of water supply systems. In North America's more arid regions, the relationship dare not be ignored.

Third, expansion of an urban roadway network, in response to increased urban travel demand, induces a reduction in urban development densities and stimulates even further travel demand and further roadway capacity expansion. Regulating the strength of this feedback loop is strategic to the long-term management of urban growth, traffic, energy use, atmospheric emissions, and water demand.

Relevance

Any effort to understand and portray the trajectories of human-environmental interactions within this continent's urbanized regions requires an explicit treatment of urban land, urban transportation, and their mutually reinforcing relationship. Failure to do so neglects one of the most significant structural relationships governing the long-term behavior of any North American urban system.

Moreover, without the inclusion of this reinforcing relationship, any model that purports to connect urban demo-economic forces to urban travel demand or to urban environmental and natural resource sectors will be badly misspecified. In this regard, the practice of regional travel demand forecasting merits special note.

Four-step travel demand forecasting models are regularly used by metropolitan planning organizations to forecast regional travel demand, compare future demand to current road capacity and recommend appropriately sized additions to regional highway capacity. These models typically do not contain a feedback loop from increased highway lane miles to decreased development densities and increased trip generation. Thus, these models regularly misrepresent the impact that added highway capacity will have on reducing congestion delay. As a result, they regularly overestimate the benefits of highway construction relative to the no-build alternative.

Of course, accurately modeling the mutually reinforcing relationship between land development and roadway capacity expansion would limit the ease with which highway investments can be justified. But, by choosing to include the second half of a reinforcing relationship critical to the dynamics of an urban system, the nation's metropolitan planning organizations could join in building this nation's institutional capacity to design sustainable development policies.

B. Objectives

Five specific objectives are set forth. These are to:

- develop a system dynamics model of urban demo-economic, land use, and transportation relationships
- calibrate the model for a generalized North American metropolitan region with representative parameters
- explore the behavior of the model in response to selected policy variables
- explore the responsiveness of the model to selected assumptions and parameterizations
- suggest, as an indication of future direction, a framework into which additional urban environmental sectors might well fit

The goal is to create a generalized multisector system dynamics model that will be operational, generally representative, instructive, and capable of simulating macro-policy effects. The expected benefit is a sharper understanding of urban dynamics and of the effectiveness of alternative policies for the management of urban land and infrastructure development.

C. Method and Model

The process for the development of a system dynamics model has six steps:

1. define the problem, state the issue
   a) explicitly state the purpose
   b) posit a reference behavior pattern
   c) develop a system diagram
2. develop and represent a thematic and dynamic organizing principle
   a) seek a dynamic organizing principle
   b) map and characterize the stock/flow hypothesis
   c) add necessary functions and parameters: make the hypothesis simulatable
3. construct a model to test the validity of the organizing principle
4. design and test policies and scenarios
5. challenge the boundaries of the model
6. make the learning available to others

Step 1: Define the problem, state the issue

From regional transportation planners to the man on the street, nearly everyone understands that new urban land development generates additional traffic and adds cumulatively to the need for more roadway capacity. Relatively few understand that the cumulative addition to roadway capacity encourages further land development, lowers developmental densities, induces both longer and more frequent travel behavior, and limits alternatives to automobile dependency. We take issue with this imbalance in understanding. To rectify the imbalance, we seek to develop an urban land-use and transportation model that captures the effect that expansion of roadway network capacity has on induced land development and travel behavior.

A reference behavior pattern represents the hypothesized trajectories of a few stocks and flows that characterize the central aspects of the system under study. A hypothesized reference behavior pattern for this study is shown in Figure 1. There one sees a graph whose x-axis marks off 25 years of time and whose y-axis measures for some prototypical urban area growth in population, urban roadway lane mile per capita, and urban land use per capita. One quickly sees that while population grows linearly, roadway capacity measured as *LaneMilesPerCapita* grows exponentially. One can also see that with minimal delay there is shown an equivalently rapid increase in urban land development measured as *UrbanLandPerCapita*. Also assumed but not shown are exponential increases in vehicular traffic, road building, land development, and municipal water use. Were we modeling air quality, it would be reasonable to assume equivalent increases in traffic-based atmospheric emissions unless offset by technical advances in emissions control. Of the various sensible reference behaviors that could be posited, all

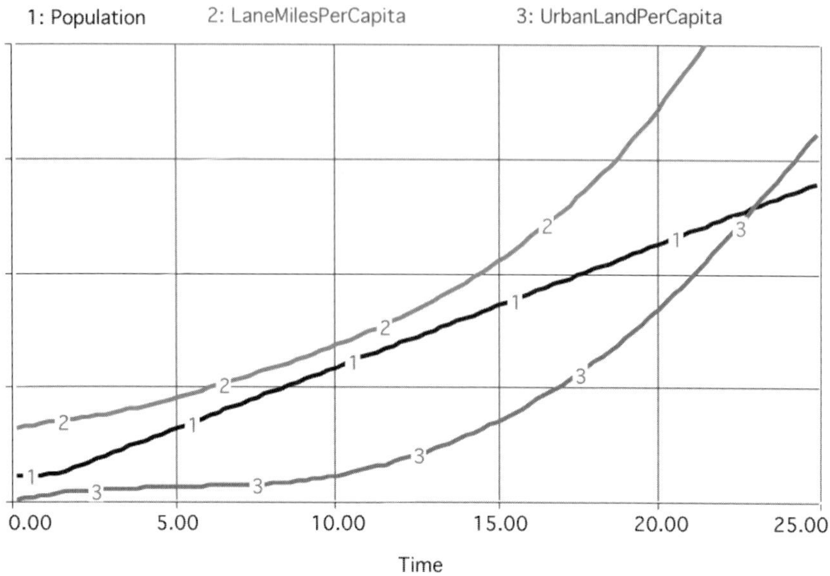

Figure 1: Hypothesized reference behaviors for urban land use and transportation interaction.

would exhibit a rate of change well in excess of either population or employment, both of which are thought to drive the dynamics of a metropolitan system.

Positing a set of reference behaviors in this manner invites a question as to why these referents would so outpace the factors assumed to be their cause. Rising household income is often advanced as the reason for urban land development expanding faster than the underlying base of population and employment. Yet there is no evidence of incomes rising exponentially during the past three decades of expanding urban land use per capita. To sort out the riddle, we consider broad inter-sector relationships and then the model's dynamic organizing principle.

A system diagram functions at a high level of generality to suggest broad relationships among sectors. Figure 2 shows a diagram defining relationships among four urban sectors—a demo-economic sector, an urban land development sector, a transportation sector, and a water sector.

The diagram may be interpreted as follows. Demographic and economic change affects rates of urban land development. The development of urban land creates traffic and increases demand for transportation facility investments. Transportation facility investments encourage urban land development. To suggest how this simple structure affects rates of natural resource use, we have included an urban water sector characteristic of an arid climate regime. Urban activities based on developed land require certain amounts of water per developed acre.

Thus the link between demo-economic growth and demand for natural resources is mediated by the interaction between land development and transportation. Because this interaction is structured as a mutually reinforcing feedback

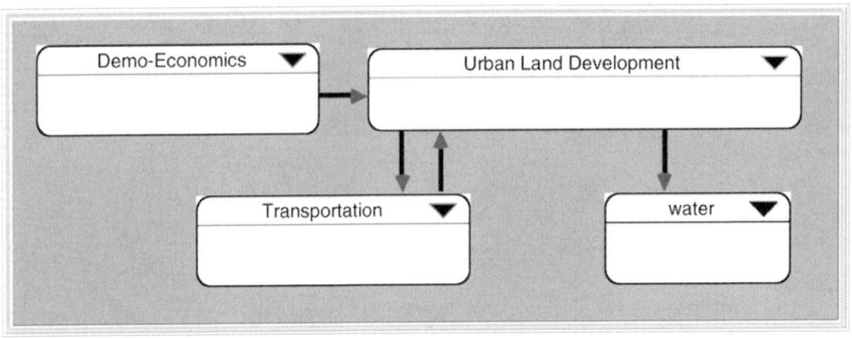

Figure 2: A system diagram among broadly defined urban sectors.

loop, it constitutes, in addition to population and employment growth, a third driving force within the urban system—an autonomous force sufficient in strength to maintain an expansion of environmental emissions and natural resource use even in the absence of any economic or demographic impulse.

Step 2: Develop and represent a thematic and dynamic organizing principle

The dynamic organizing principle for the model lies in the way demo-economically driven land development generates travel demand and the need for added roadway capacity, the construction of which, through a specific feedback effect, lowers land-use densities and induces an even more rapid rate of land development per capita or per job. This principle is the basis for hypotheses about the rapid rates of growth in travel demand, road construction and urban land development relative to rates of population and employment growth.

We can begin to explore this dynamic with a system map. A system map serves as a diagram of stock and flow relationships. These are best considered sector by sector. The demo-economic sector in Figure 3a is a systems interpretation of an economic base model with employment and population multipliers. Employment growth is understood to drive the system.

Employment is divided into two parts—basic and nonbasic. Basic employment refers to all employment dedicated to the production of goods and services for export beyond the metropolitan region. The income generated through basic employment is enjoyed by the families that support nonbasic employment—employment dedicated to the production of goods and services for local consumption. A multiplier relationship exists between the two so that change in basic employment leads to amplified but proportionate change in nonbasic and in total employment. Similarly, a change in employment induces a proportionate change in population. Otherwise, population is presumed to remain stagnant with intrinsic growth exactly offset by intrinsic decline. This last assumption simplifies

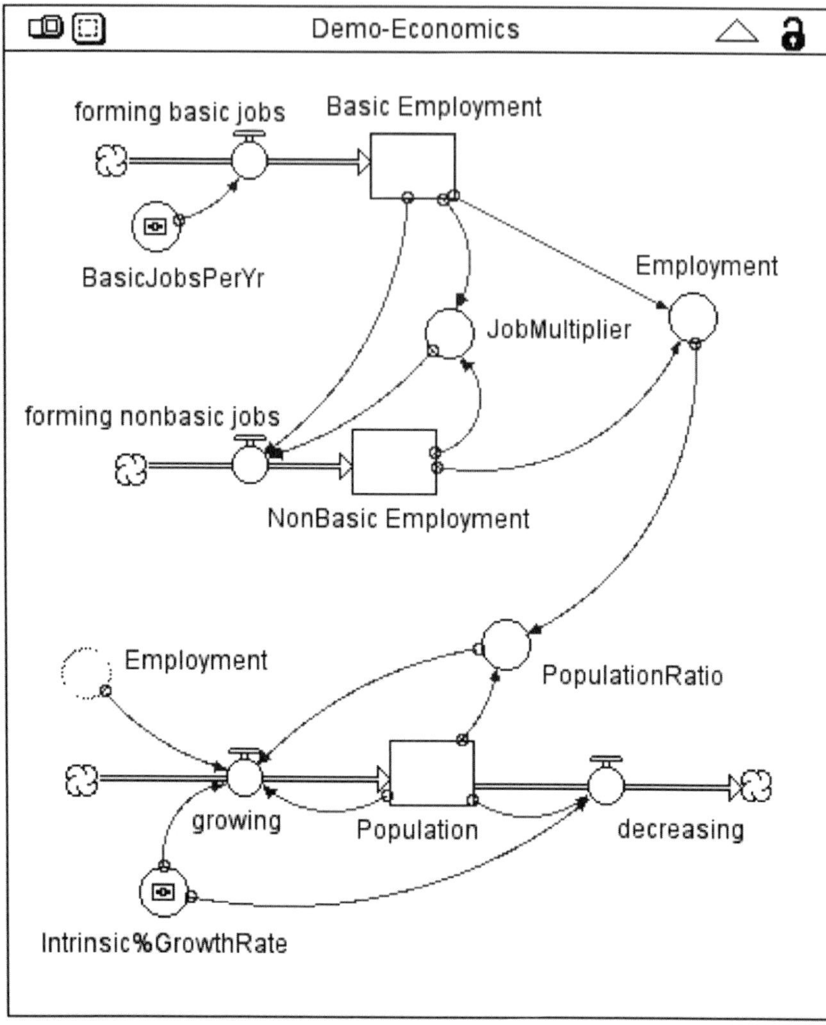

Figure 3a: Stock/flow diagram: demo-economic sector.

the demo-economic sector by making population change a sole function of basic employment change.

A modest difference may be noted between this representation and the classic representation of the economic base model. There, the economic base multiplier and the population-to-employment multiplier are exogenously given constants. Here they are calculated internally on the basis of the prior period's outcomes. As a result, they vary modestly as the system's dynamics unfold. Arguments may be readily advanced to suggest that this is the more reasonable treatment. Otherwise, basic employment is set to grow at a rate set by a slider that can be varied between zero and 100 basic jobs per year. The default setting is 50 basic jobs per year. Initial conditions and exact equations are specified in Appendix A.

Next consider the sector for urban land development illustrated in Figure 3b. The jobs and people simulated in the demo-economic sector create a need for urban land development. Initially, population growth, the number of persons per household, and the density of residential land development determine the annual amount of residential development. This is added to the annual amount of commercial land development due to job growth and the average acreage per job. Subsequently, rates of land development are controlled by the added effects that expanded roadway capacity has on developmental densities *(∂ResidDensity:∂Roads and ∂ComDensity:∂Roads)* and by any development densification policies that might be enacted.

Figure 3b: Stock/flow diagram: urban land development sector.

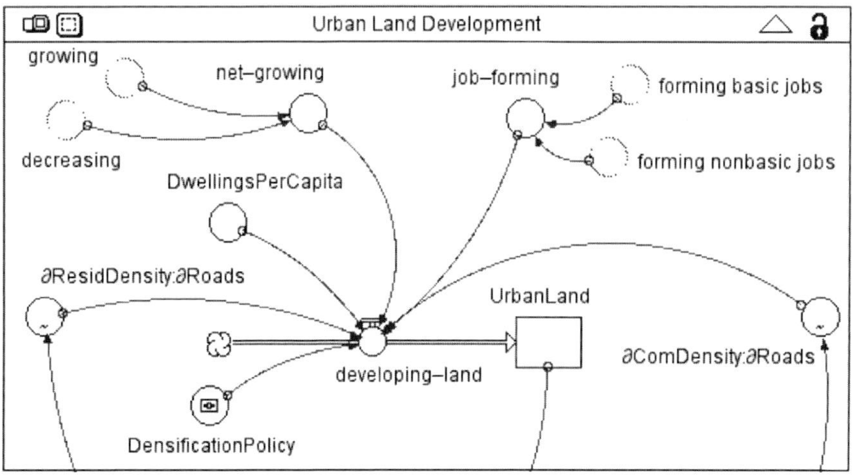

Figure 3c: Stock/flow diagram: transportation sector.

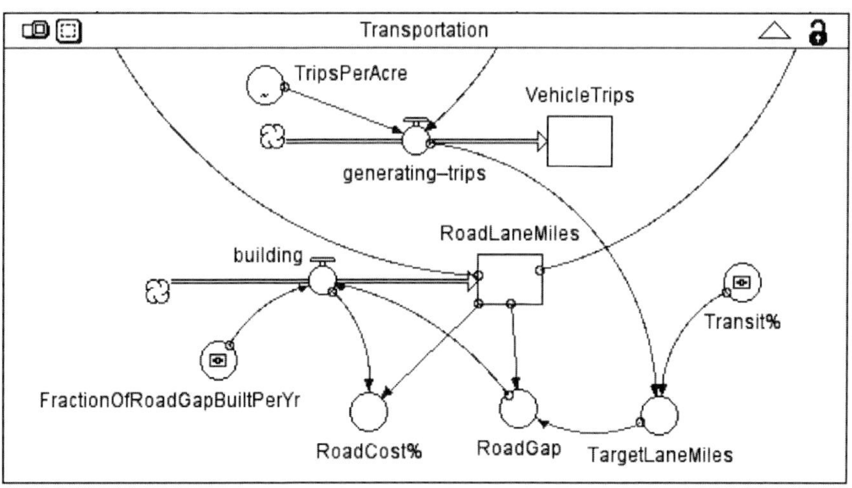

While some find it curious, a standard practice in transportation planning is not to argue that people generate traffic but to assign the traffic generation function to land and to various land uses. Developed land generates traffic. Traffic engineers assign each land use a standard traffic generation rate per acre. We follow the same practice here.

Growing travel demand creates a gap between current and needed roadway capacity. The gap is filled through road construction. Despite efforts to fill the gap, it expands monotonically. As more roads are built, more remote locations become accessible, the supply of developable land is expanded and commercial and residential densities slowly decline. This last impact, declining densities, has complex economic, behavioral and institutional antecedents, but it rarely occurs without transportation facility investments and a consequent increase in accessible land.

The diagram in Figure 3c represents this logic stylistically. *TripsPerAcre* are defined as a modestly increasing function of time. This captures long-term effects of increased automobile ownership and increased incomes on travel behavior. *TripsPerAcre* per day multiplied by the acres of developed land (*UrbanLand*) yields a daily flow of person trips (*trip generation*). These, as modified by the variable proportion of trips made by transit, determine a desired level of roadway capacity (*TargetLaneMiles*). The target is continually compared to existing roadway capacity (*RoadLaneMiles*) to determine the current road construction gap (*RoadGap*). Expanding roadway capacity is a goal-seeking process that strives to catch up with an ever-increasing gap. Road capacity is expanded each year to cover a specified fraction of the current road capacity gap. The *FractionOfRoadGapBuiltPerYear* can be adjusted to simulate a more aggressive or less aggressive road building policy.

The physical extent of the roadway network (measured in *RoadLaneMiles*) influences the densities at which newly developed commercial and residential land is settled. As population and job growth proceed, land developed at lower and lower densities accumulates quickly, requires longer and more frequent person trips, and thus generates a greater travel demand.

The impacts of more rapid land development on growth in vehicular energy consumption and atmospheric emissions, though not shown, are intuitively obvious. But we have included and can show a highly stylistic water consumption sector. This is shown in Figure 3d.

Figure 3d assumes a climate that requires irrigation of lawns—typically a climate where water is scarce. As residential and commercial densities decline, more water is needed per developed acre (*GalPerAcrePerYear*). This growing rate multiplied by the ever-increasing acres of urban land yields an estimate of annual water consumption. This may be modified by a water conservation factor that presumes a user-defined percentage reduction in consumption due to conservation practices.

The four sectors are linked together in a multisector system map (Figure 3e).

Worth noting here are the connections between sectors that are not shown in the prior sector maps. Thin arrows drawn from an element in one sector to an element

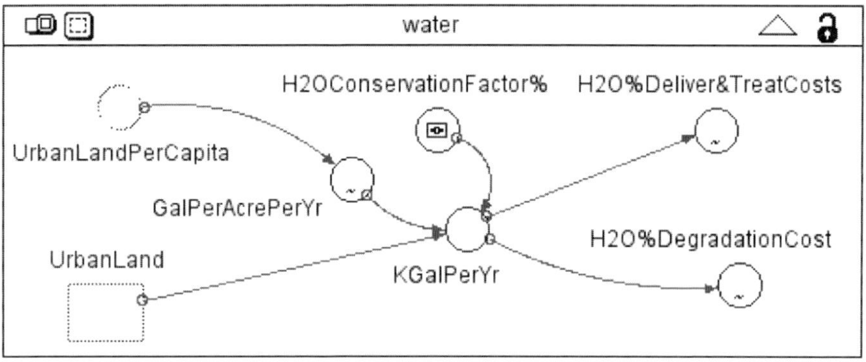

Figure 3d: Stock/flow diagram: municipal and industrial water sector.

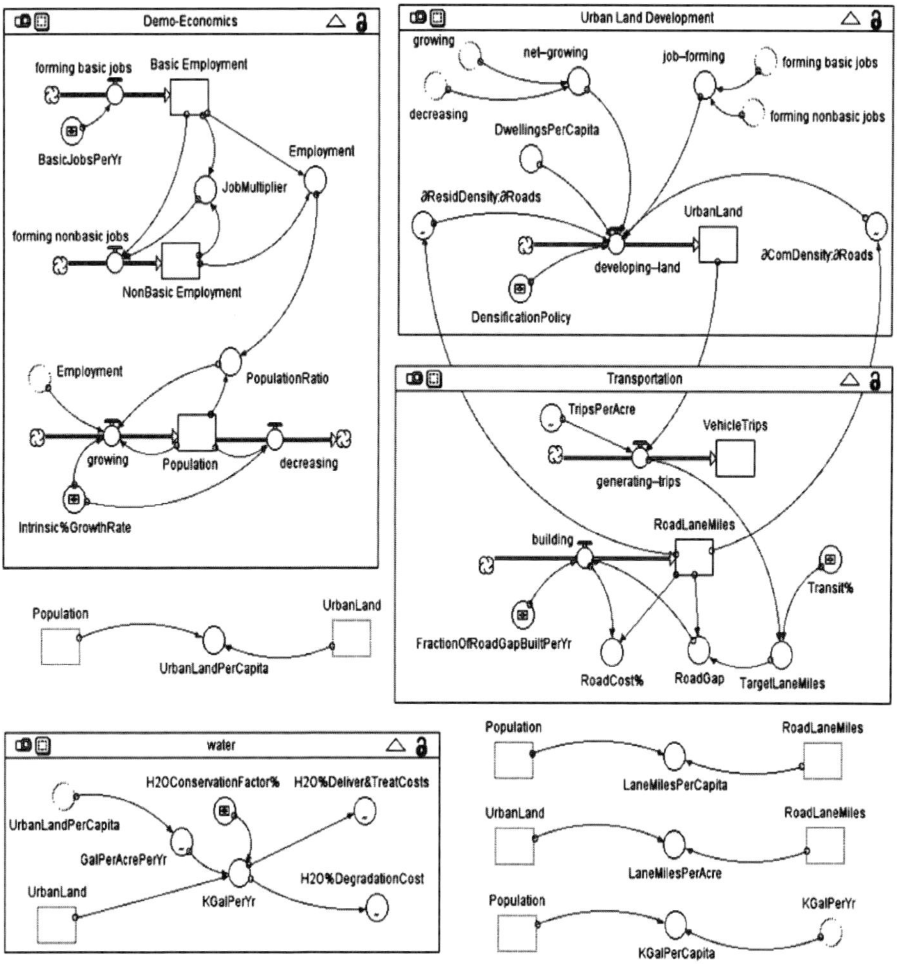

Figure 3e: Stock/flow diagram for a model of land use and transportation.

of a different sector illustrate the indicated connections. Ghosted images of an element established in one sector but replicated in another sector also show connections between sectors. Operations outside of the four sector boxes are for the calculation of rates useful elsewhere in the model. By tracing connections among sectors, one can bring the logic of the entire model into focus.

The model's logic is direct and straightforward. People and jobs generated in the demo-economic sector drive the flow of developing land that propels trip generation and road building. The accumulation of *RoadLaneMiles* induces developmental density declines that spur on the flow of developing land. Implications of accumulating *UrbanLand* and *UrbanLandPerCapita* for disproportionate growth in water use and infrastructure costs are traced out in the water sector.

Not shown but easily imagined are the implications of a disproportionate increase in trip generation for vehicular energy consumption. Also not shown but easily imagined are the implications for congestion delay and vehicular emissions. These deteriorate because the increase in *LaneMilesPerCapita* cannot possibly keep up with the even more rapid increase in trip generation. In spite of a vigorous road construction schedule, the *RoadGap*—a decent surrogate for congestion delay—will grow disproportionately, as do all things swept up in this vicious cycle.

Step 3: Construct a model to test the validity of the organizing principle

The third step adds functions and parameters to make the model operational. Functions and parameters are detailed in Appendix A. Highlights will be discussed here with particular attention to three functions that should be treated with special scrutiny.

Consider first *TripsPerAcre*. This parameter is not a constant but a variable function of time changing from four trips per acre in year zero to 16 trips per acre in year 25. It transforms growth in *UrbanLand* to even more rapid growth in *trip generation*. There is a general consensus that trips-per-acre is a monotonically increasing function of time. The extent of the change over longer periods of time is still a point of active debate. Alterations in *TripsPerAcre* as a function of time will affect the trajectory of trip generation through time. Nonetheless, the general pattern of system behavior postulated earlier as the system's reference behavior will hold whenever *TripsPerAcre* as a function of time is monotonically increasing.

Next consider the impact of *RoadLaneMiles* on developmental densities. Expanding *RoadLaneMiles* lowers land development densities. This relationship is captured in the two graphical functions, *∂ResidDensity:∂Roads* and *∂ComDensity:∂Roads*. *∂ResidDensity:∂Roads* makes residential densities a function of regional road lane miles. It varies between seven units per acre when there are only 100 miles to the road network at the beginning of the simulation to just over one unit per acre when there are 1,100 miles of roadways near the end of a more road-intensive simulation. *∂ComDensity:∂Roads* does the same for commercial densities. It varies between nine units per acre when there are only 100 miles of road to just over one unit per acre for the more road-intensive simulations. Both functions imply negative long-term elasticities of developmental densities with respect to road lane miles below 0.10. Stratham (2000) reports elasticities that are more than double this value. Our specification of these two relationships seems to be conservative.

While there is a general consensus that developmental densities decline with the long-term expansion of regional roadway capacity, the best evidence in support of our current specifications is indirect. This evidence is drawn from the literature on the impacts of highway capacity expansion on induced travel demand.

Induced travel demand is created pursuant to the expansion of roadway capacity. It is due, in part, to changes in travel behavior stimulated by new capacity (e.g., switching from other routes, other times of day, and other modes of travel, decreasing vehicle occupancy, increasing trip frequency, etc.). But it is also due to the relocation by households and commercial activities to the more remote sites where recent development was promoted by roadway construction.

The Standing Committee on Trunk Road Assessment (1994) advanced the saliency of this line of inquiry when it affirmed research suggesting that there was a one percent increase in induced traffic for every one percent increase in trunk road capacity. A well-cited essay by Hanson and Huang (1997) states that within four years of construction, a 0.9 percent increase in vehicle traffic occurred for every one percent increase in California county highway lane miles. A recent review by Stratham (2000) reports state-wide and metro-area induced travel elasticities with respect to highway lane miles ranging from 0.6 to 0.90 with a mean value of 0.83.

The simulations produced by the model presented here have, under default settings, an elasticity of *PersonTrips* with respect to *RoadLaneMiles* in the range of 1.00–1.10. This result suggests that while the functions considered above are not grossly in error they would benefit from modest adjustment.

Finally, consider the question of the presupposed reference behavior and whether simulations with the current model confirm presuppositions. This takes us to the part of the system dynamics software called the cockpit. Here the user is presented with a series of policy variables arranged as sliders. Each can be set to a desired level within a prespecified range. Figure 4 shows how this is arranged.

Sliders permit manipulation of *FractionOfRoadGapBuiltPerYear* (0.33), *Transit%* (3%), *DensificationPolicy* (0%), *Intrinsic%GrowthRate* (1.4%), *Basic JobsPerYear* (50), and *H2OConservationFactor%* (0%). Figure 4 shows default settings. With sliders in their default positions, observations can be made on simulation outcomes with respect to *Population*, *LaneMilesPerCapita,* and *UrbanLandPerCapita*. These are shown in Figure 5a.

Figure 5a shows *Population* growing linearly while *LaneMilesPerCapita* grows exponentially. *UrbanLandPerCapita* grows exponentially, too, but at first more slowly than *LaneMilesPerCapita* and later with increasing speed. Once equilibrated, the elasticity of *UrbanLand* to *LaneMiles* ranges between 0.5–0.7.

This pattern is complemented by the trajectories of other system stocks and flows. While *Population* increases by 310 percent, *UrbanLand* increases exponentially by 8,200 percent. Daily *trip generation* grows exponentially by 3,300

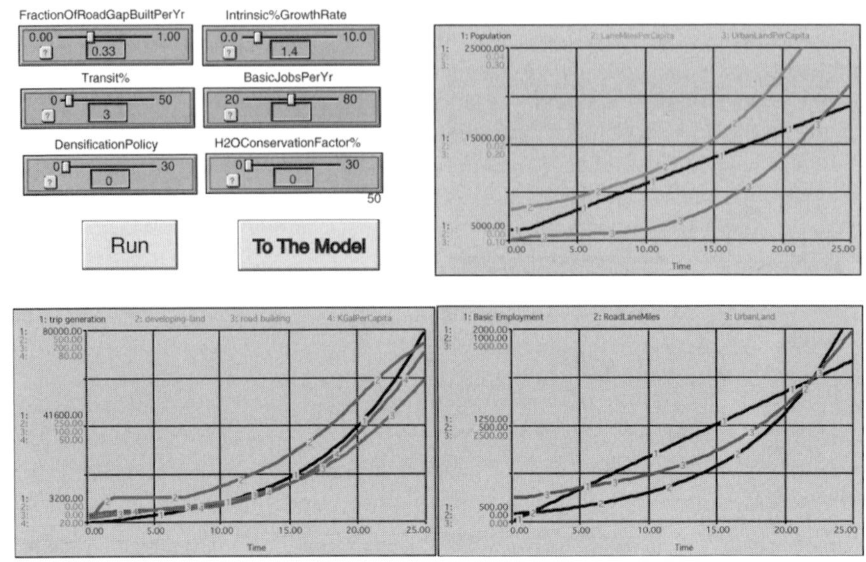

Figure 4: A view of the software's cockpit level.

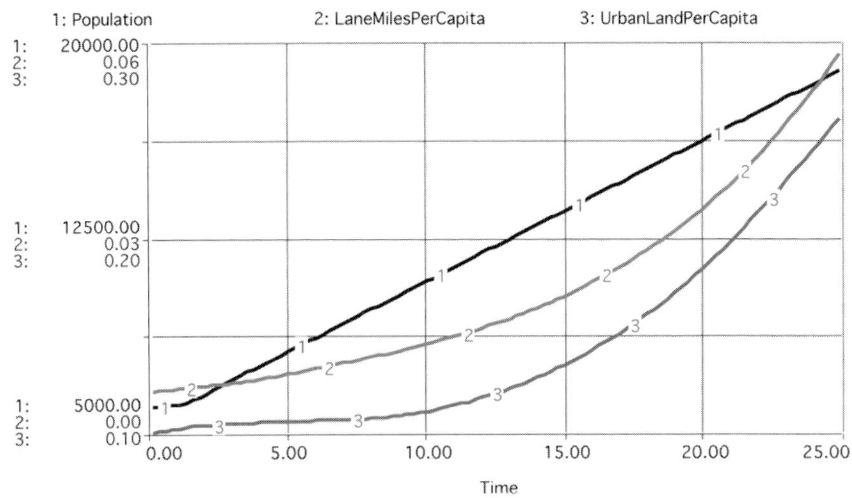

Figure 5a: A simulation outcome for reference variables.

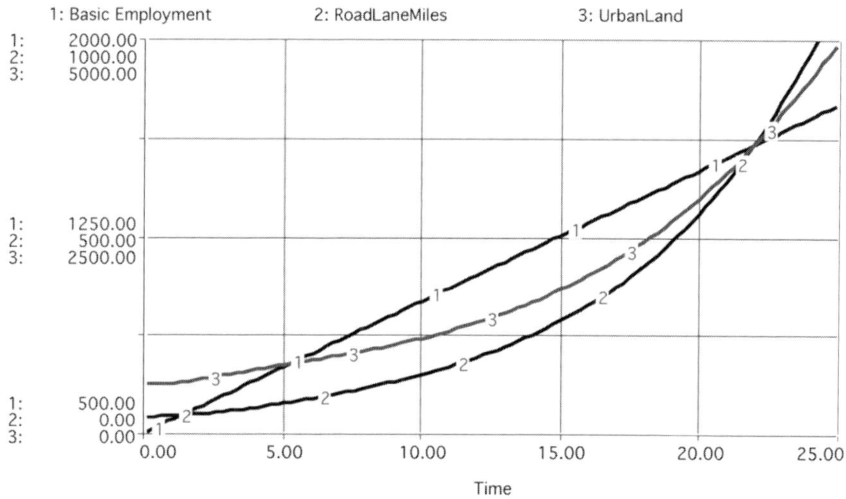

Figure 5b: A simulation outcome for additional reference variables.

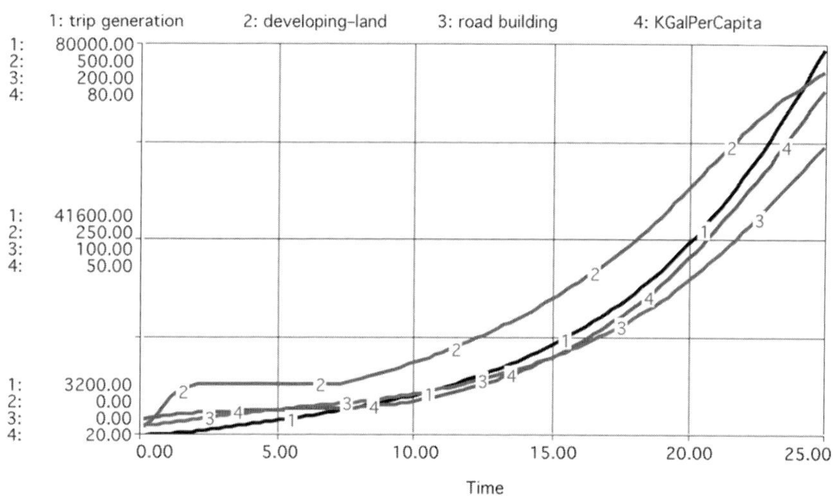

Figure 5c: A simulation outcome for further reference variables.

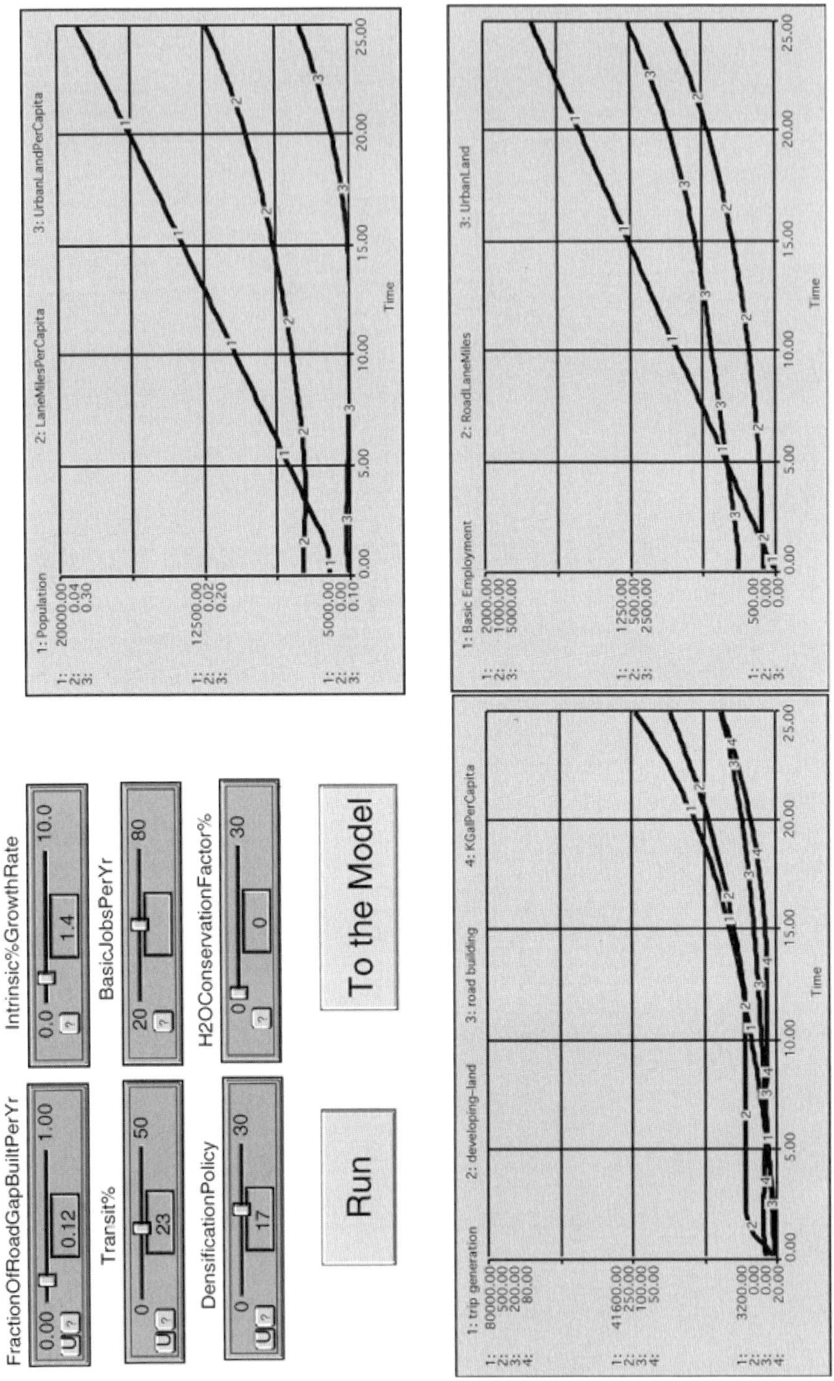

Figure 6: Simulation outcomes for a dampened feedback scenario.

percent. *RoadLaneMiles* increases exponentially by 3,000 percent. Despite the intense road building activity, the *RoadGap* increases exponentially by 4,400 percent. The dynamic organizing principle underlying the reference behavior pattern is amply confirmed. These results are seen graphically in Figures 5b and 5c below.

Step 4: Design and test policies and scenarios

Simulation results under default settings show an aggressive pattern of urban fiscal and environmental resource use. Once the dynamic organizing principle is understood, moderating this aggressive pattern is relatively straightforward. Moderation is achieved by weakening the reinforcing feedback effect between roadway capacity expansion, reduced developmental densities, and increased vehicle trip generation. To do this, we use land-use controls to increase densities, mass transit to reduce the vehicle travel demand, and the *FractionOfRoadGapBuiltPerYear* to slow road construction.

Figure 6 shows the trajectories of various stocks and flows under this dampened feedback scenario. The *FractionOfRoadGapBuiltPerYear* has been lowered from 0.33 to 0.12. The *Transit%* has been raised from 3 to 23 percent. The effect of *DensificationPolicy* on land development densities has been changed from 0 to 17 percent.

Growth in *UrbanLandPerCapita* is reduced by 50 percent. Growth in *LaneMilesPerCapita* slows from a seven-fold increase to a three-fold increase. Daily *trip generation* is reduced by 50 percent. *RoadLaneMiles* built over the 25-year period are reduced by 67 percent. The surface area of the simulated city is half as large. *PersonTrips* are reduced by a third. Automobile-based person trips are half as numerous. Automobile-based *PersonTripsPerLaneMile* are up by one-third, but with the city's surface area reduced by one-half, these trips are equivalently shorter. After 25 years, the *RoadGap* is reduced by one-half. This is despite the reduced intensity of road building.

All this suggests that roadway congestion and vehicle traffic delay might actually be less under the dampened feedback scenario even though road building is much reduced. Moreover, our surrogate for natural resource use, municipal and industrial water use per capita, is reduced by more than 50 percent, and this without any direct water conservation measures being taken.

One's assessment of these differences will vary by personal preference. Nonetheless, the dampened feedback scenario does suggest numerous fiscal, environmental, and natural resource benefits together with likely improvements in congestion delay.

Step 5: Challenge the boundaries of the model

The contrast between the prior two scenarios is striking. The contrast is attributed to a feedback loop, the effects of which are strategically dampened by a set of three policy variables. Are there alternative ways to explain the contrast between scenarios? Could this contrast be a result of the way the model is bounded? Are there relationships between the sectors that are not now considered? Are there other sectors not now considered that, if included, could alter the dynamic demonstrated above?

Logic dictates that, to fundamentally alter the model's current dynamic, a counteracting feedback loop would be needed. This would autonomously dampen the dynamic that now defines the model's behavior. There are three ways in which this might occur.

First, a counteracting feedback effect might be imagined that would lessen the impact that added highway capacity has on decreasing developmental densities. This might occur with the autonomous emergence of suburban poly-nucleations each with increased commercial and residential densities. While this would dampen the model's current dynamic, it most likely would not alter or overcome its dynamic organizing principle.

Second, the dispersal of employment opportunities implicit in commercial density declines might promote a shorter journey-to-work and a shorter shopping trip length. Stratham (2000) found evidence of this to be negligibly small.

Third, a counteracting feedback effect might be imagined that would operate on the propensity to form basic jobs. Are there mechanisms that might produce this effect with sufficient force to alter the behavioral dynamic of the model's current structure? One possibility comes to mind. The obvious disfunctionality of the unmitigated growth scenario could over a sufficiently long period of time create urban environmental and fiscal conditions unfavorable to continued growth in employment and population. High taxes, long commutes, congestion delays, air pollution, and high housing costs eventually discourage business expansion and drive off amenity-oriented households. This dynamic, if included in our model, would eventually drive exponential expansion into a pattern of overshoot-and-collapse. Only with the most patient pattern of public and private cooperation for urban growth management, could an exponential impulse be deflected toward a sustainable growth trajectory.

The model's dynamic behavior is obviously due to the impact of added roadway capacity on decreasing developmental densities. Other than the *DensificationPolicy*, it is hard to imagine an effective counteracting feedback mechanism that would offset this effect. Would a price mechanism do the job? The value of increased accessibility provided by added roadway capacity is readily capitalized into the price of nearby land. But this price increment is gained in large part by the collective diminution of values throughout the rest of the regional land market. The land market price mechanism modulates but does not reverse the tendency that an increased supply of accessible land has on encouraging greater consumption through lower overall prices.

The model does not have a fiscal sector. Would its inclusion change things? The increased public expense of an exponential expansion in urban land could perhaps be an effective break on the model's self-reinforcing dynamic. But at present, state, county, and municipal fiscal policies spread most public costs for new development among the occupants of both existing and newly developed sites. Thus, a principal effect of roadway expansion is to underwrite nearby site values at the expense of the public as a whole. An equitable taxing mechanism would have a dampening effect on feedback dynamics, but this potential effect is mostly neutralized by the way highways and other urban infrastructure expansions are currently financed.

In addition to the direct modification of our model's current dynamic principle, one could imagine an exogenous (extra-regional) deterioration in conditions favoring continued employment and population growth. Either this situation or some combination of internal and external economic deterioration could modulate or, with enough time, reverse the dynamic shown above.

Otherwise, one may not regard the behavior generated by our model to be an artifact of an incomplete specification of the larger urban system. On the contrary, it appears that a timely recognition of the reciprocally reinforcing relationships so fundamental to metropolitan dynamics would readily facilitate an articulation of locally sustainable urban development policies.

Step 6: Make the learning available to others

The STELLA model presented here can be downloaded from the following website: <http://www.geog.utah.edu/~pcemmi/systemmodel>. Depending upon your platform, download either the Stuffit document named <UrbSys1Mac.sit> or the ZipIt Document named <UrbSys1PC.exe>. Extract the system dynamic model from its compressed form to get the file named <UrbSys1.stm>. The extracted STELLA model may be run on a save-disabled version of Runtime STELLA version 7.0 available without charge at <http://www.hpsinc.com/Register /STELLA DemoKit.asp>

To navigate around the model, use the up/down triangles in the upper-left corner to switch among program levels. The upper-most level contains the graphs and control sliders. Set these as you wish and click on the "run" button. The middle level contains the system map with sector details. Go here to review or modify the structure of the model. The lower level contains the 47 simultaneous difference equations that make up the model.

D. Summary and Conclusions

Within North American metropolitan regions, we have observed over the past three decades rates of growth in urban land development, vehicular traffic, and roadway construction that consistently outpace associated rates of growth in either population or employment. The observation merits an explanation. One way to proceed is to design a system dynamics model that replicates this observed pattern of behavior and then to seek an explanation within the structure of relationships among the model's elements.

System dynamics modeling is grounded in systems thinking. This involves at least four operations: (1) laying out the mechanisms one suspects are responsible for observed behaviors, (2) representing these in an operational systems model, (3) using the model to explore the system's range of responses, and (4) questioning the soundness of the model's boundary assumptions.

Mediating between urban demo-economic forces that drive metropolitan systems and the use of urban fiscal and environmental resources is a nexus of mutually reinforcing relationships. These govern the interaction between urban land development, trip generation, and roadway construction. In systems terms, this nexus is a goal-seeking process nested within a self-reinforcing feedback loop. The result of this structure is incessantly more fervent activity in pursuit of an ever-receding goal. Increasingly more miles of roadway construction induce developmental density declines and increased vehicular traffic. These produce an ever-expanding gap between desired and actual roadway capacity.

Our model of this dynamic shows how linear expansion in employment and population yields an exponential expansion in urban land development, trip generation, and road building. It suggests that urban land-based water consumption and urban travel-based energy consumption expand exponentially as well.

Our model also provides an experimental environment within which to explore the outcomes of selected policy manipulations and the effects of certain scenario designs. A dampened feedback scenario seeks to moderate the strength of the feedback dynamic between land development and road building. It does so by lowering the intensity of road construction, expanding transit service, and increasing developmental densities. The combined policies substantially reduce the otherwise exponential rates of increase in urban land consumption per capita and road miles per capita. With that, a series of other arguably beneficial developments ensue.

While still a prototype, the validity of the model is confirmed in three ways. First, the simulations produced by the model have an elasticity of *PersonTrips* with respect to *RoadLaneMiles* in the range of 1.00 to 1.10—a result not badly out of line with other empirical studies. Second, the internal structure of the model conforms to fundamental understandings of how cities work, is internally consistent in its units of measurement, produces patterns of behavior congruent with hypothesized behaviors, and offers a satisfactory explanation for the behavior it seeks to replicate. Third, the boundary assumptions employed by the model do not appear to be unreasonable, nor do they offer a plausible alternative for why the model behaves as it does.

The next steps will be to transform the graphical relationships used in the model into mathematical functions and to test for the sensitivity of the model's reference behavior to variation in each function's characteristic parameters. If the parametric specification for any mathematical function is found to greatly influence the model's reference behavior, then we will need to do an extensive search of the empirical literature for evidence of the numerical values found for that parametric specification.

However, we strongly suspect that the model's behavior is governed by its general structure. Numerical variation across likely ranges of parametric specification will only modify the time frame within which the reference behavior unfolds and not its general pattern of behavior. In short, we suspect that any necessary empirical correction to the model's functional relationships will affect only the amount of time required to demonstrate the presupposed reference behavior and not the behavior itself.

Because of its reciprocally reinforcing feedback structure, we suspect the model will reliably transform linear growth in demo-economic factors into exponential growth in urban environmental factors regardless of the numerical variations likely to be found in the empirical literature on the mathematical functions used to link its various parts one to the other. By inference, an understanding of metropolitan dynamics rests largely upon a comprehension of the general structure of relationships and not on precise empirical estimates of induced land use and travel demand elasticities. While the model does rely upon empirical investigations to identify its requisite parts, it also tolerates considerable ambiguities in empirically grounded elasticity estimates.

While the work presented here is in several ways preliminary, current results do suggest that the present approach can lead to clear and possibly compelling insights into the subject we explore. In particular, it suggests that a mutually reinforcing relationship involving urban land development, trip generation, and road construction explains why urban land develops at rates well in excess of demographic or economic growth. It explains how our history of incessantly more fervent road building undertaken in pursuit of the ever-receding goal of congestion relief has induced density declines, increased vehicular traffic, and created an ever-expanding gap between desired and actual roadway capacity. It explains how this history has left us with a legacy of increasing infrastructure expenditures, growing roadway congestion, unabated vehicular energy use and a steady stream of local real estate booms in increasingly spread-out suburban property developments. It defines the generalized long-term consequences of continuing in this manner. It identifies a three-fold policy design for mitigating the untoward consequences of present developmental patterns. It offers an opportunity to experiment with alternative policy initiatives and invites members of this epistemic community to learn and act on the understandings they may gain.

References

Forrester, Jay W. 1961. *Industrial dynamics*. Waltham, Mass.: Pegasus Communications.

———. 1969. *Urban dynamics*. Waltham, Mass.: Pegasus Communications.

———. 1971. *World dynamics*. Waltham, Mass.: Pegasus Communications.

Friedmann, John. 1987. *Planning in the public domain: From knowledge to action*. Princeton: Princeton University Press.

Hansen, M. and Y. Huang. 1997. Road supply and traffic in urban areas: A panel study. *Transportation Research. Part A: Policy and Practice* 31 (3):205–218.

Lane, David C. 2001. Rerum cognoscere causus: Part I–How do the ideas of system dynamics relate to traditional social theories and the voluntarism/determinism debate? *System Dynamics Review* 17 (2):97–118.

Richmond, Barry. 1994. System Dynamics/System Thinking. Presented at the International System Dynamics Conference, Sterling, Scotland, July 25–29. Available at URL: <http://www.hps-inc.com/hps_resources.htm>.

Standing Committee on Trunk Road Assessment. 1994. *Trunk roads and the generation of traffic*. London: HMSO.

Strathman, J. G., K. J. Dueker, T. Sanchez, J. Zhang, and A. Riis. 2000. Analysis of induced travel in the 1995 NPTS. Presented at the Association of Collegiate Schools of Planning Conference, Atlanta, November 5–10. Available at: <http://www.upa.pdx.edu/CUS/PUBS/PDFs/PR113.pdf>.

Appendix A: The Model's Equations

Demo-economics

Basic_Employment(t) = Basic_Employment(t - dt) + (forming_basic_jobs) * dt
INIT Basic_Employment = 500
forming_basic_jobs = BasicJobsPerYr
NonBasic_Employment(t) = NonBasic_Employment(t - dt) + (forming_nonbasic_jobs) * dt
INIT NonBasic_Employment = 2000
forming_nonbasic_jobs = JobMultiplier*(Basic_Employment-DELAY(Basic_Employment,1))/(1-JobMultiplier)
Population(t) = Population(t - dt) + (growing - decreasing) * dt
INIT Population = 6000
growing = PopulationRatio*(Employment-DELAY(Employment,1)+(Population*Intrinsic%GrowthRate/100)
decreasing = Population*Intrinsic%GrowthRate/100
BasicJobsPerYr = 50
Employment = Basic_Employment+NonBasic_Employment
Intrinsic%GrowthRate = 1.4
JobMultiplier = NonBasic_Employment/(Basic_Employment+ NonBasic_Employment)
PopulationRatio = Population/Employment

Transportation

PersonTrips(t) = PersonTrips(t - dt) + (trip_generation) * dt
INIT PersonTrips = 15000
trip_generation = UrbanLand*TripsPerAcre
RoadLaneMiles(t) = RoadLaneMiles(t - dt) + (road_building) * dt
INIT RoadLaneMiles = 37
road_building = RoadGap*FractionOfRoadGapBuiltPerYr
FractionOfRoadGapBuiltPerYr = .33
RoadCost% = (1.5*RoadLaneMiles+20*road_building)/INIT(RoadLaneMiles)
RoadGap = TargetLaneMiles-RoadLaneMiles
TargetLaneMiles = .02*trip_generation*(100-.7*Transit%)/100
Transit% = 3
TripsPerAcre = GRAPH(time)
(0.00, 4.00), (2.08, 5.00), (4.17, 6.00), (6.25, 7.00), (8.33, 8.00), (10.4, 9.00), (12.5, 10.0), (14.6, 11.0), (16.7, 12.0), (18.8, 13.0), (20.8, 14.0), (22.9, 15.0), (25.0, 16.0)

Urban land development

UrbanLand(t) = UrbanLand(t - dt) + (developing–land) * dt
INIT UrbanLand = 600

developing–land =
((1.1*net–growing*DwellingsPerCapita*∂ResidDensity:∂Roads)+(job–formin
g*∂ComDensity:∂Roads))*(100-DensificationPolicy)/100
DensificationPolicy = 0
DwellingsPerCapita = .4
job–forming = forming_nonbasic_jobs+forming_basic_jobs
net–growing = growing-decreasing
∂ComDensity:∂Roads = GRAPH(RoadLaneMiles)
(100, 0.115), (200, 0.235), (300, 0.345), (400, 0.455), (500, 0.585), (600, 0.705), (700, 0.805), (800, 0.885), (900, 0.925), (1000, 0.95), (1100, 0.97)
∂ResidDensity:∂Roads = GRAPH(RoadLaneMiles)
(100, 0.141), (200, 0.28), (300, 0.42), (400, 0.528), (500, 0.618), (600, 0.699), (700, 0.766), (800, 0.825), (900, 0.879), (1000, 0.933), (1100, 0.978)

Water

H2OConservationFactor% = 0
KGalPerYr = UrbanLand*GalPerAcrePerYr*((100-H2OConservationFactor%)/100)/1000
GalPerAcrePerYr = GRAPH(UrbanLandPerCapita)
(0.00, 180000), (0.1, 220000), (0.2, 262000), (0.3, 290000), (0.4, 316000), (0.5, 340000), (0.6, 364000), (0.7, 384000), (0.8, 402000), (0.9, 412000), (1, 422000)
H2O%DegradationCost = GRAPH(KGalPerYr)
(1.5e+08, 0.05), (3.5e+08, 0.455), (5.4e+08, 0.945), (7.4e+08, 1.89), (9.3e+08, 3.05), (1.1e+09, 3.96), (1.3e+09, 4.69), (1.5e+09, 5.22), (1.7e+09, 5.64), (1.9e+09, 5.81), (2.1e+09, 6.02), (2.3e+09, 6.09), (2.5e+09, 6.20)
H2O%Deliver&TreatCosts = GRAPH(KGalPerYr)
(1.5e+08, 0.95), (3.5e+08, 0.95), (5.4e+08, 4.00), (7.4e+08, 4.00), (9.3e+08, 4.00), (1.1e+09, 7.00), (1.3e+09, 7.00), (1.5e+09, 7.00), (1.7e+09, 7.00), (1.9e+09, 9.05), (2.1e+09, 9.05), (2.3e+09, 9.05), (2.5e+09, 9.05)

Not in a sector

KGalPerCapita = KGalPerYr/Population
LaneMilesPerAcre = RoadLaneMiles/UrbanLand
LaneMilesPerCapita = RoadLaneMiles/Population
UrbanLandPerCapita = UrbanLand/Population
∂:∂ = (∂trips/PersonTrips)/(∂miles/RoadLaneMiles)
∂miles = RoadLaneMiles-DELAY(RoadLaneMiles,1,.01)
∂trips = PersonTrips-DELAY(PersonTrips,1,1)

Reexamining the Geography of the Urban Labor Market: A Case Study of the San Francisco Bay Area

Qing Shen and Mizuki Kawabata

1. Introduction

The geography of the urban labor market is formed primarily by the juxtaposition of three factors: the spatial distribution of jobs, the spatial distribution of workers, and the transportation system that connects jobs with workers. These factors jointly determine who has what level of employment accessibility in a metropolitan area. Understanding the geography of the urban labor market is essential for identifying spatially disadvantaged population groups, as well as for making effective policies to help them access economic opportunities. In the United States, it has long been recognized that the combination of employment decentralization, housing market segregation, and ineffectiveness of public transportation in serving auto-driven low-density development has created a major spatial barrier to low-income people's work participation. The issue has become a major concern for policy makers as well as urban researchers since the welfare reform of the late-1990s (Lacombe 1997, Blumenberg and Ong 1998, Rich and Coughlin 1998, Wachs and Taylor 1998). Because the new legislation enacted by Congress requires all adult welfare recipients to obtain jobs within two years of receiving cash assistance, improving accessibility of employment opportunities for people making the transition from welfare to work has emerged as a pressing task.

More generally speaking, understanding the geography of the urban labor market is essential for analyzing, modeling, and planning the city. Locations of employment and labor force, and mobility provided by transportation, are among the most important determinants of the spatial pattern of development in a metropolitan area. They are, therefore, basic variables in virtually every operational model for urban land-use and transportation planning (Lowry 1964, Wilson 1974, Batty 1976, Putman 1983, Landis 1994).[1] Knowledge of the geography of the

[1] See Wegener (1994) for a concise description and insightful review of operational urban models that represent the state of the art.

urban labor market helps planners make conditional predictions of future growth and land-use changes, which in turn provide a basis for assessing the likely social and environmental impacts. Information about desirable residential locations for workers of various socioeconomic and mobility characteristics is a prerequisite for policy making in the areas of low-income housing, infrastructure, public transportation, and social services.

In this paper, we raise two fundamental questions about the geography of the urban labor market that have not been sufficiently addressed in the literature. First, how are employment opportunities for *job seekers*—people who are currently seeking employment—spatially distributed in U.S. metropolitan areas? Second, what are the patterns of variations among job seekers in employment accessibility? We approach these questions by undertaking a case study of the San Francisco Bay Area. Applying the analytical framework developed by Shen (2001) in a recent study of the Boston metropolitan area, we first estimate *job openings*, which are employment opportunities currently available for job seekers, and examined their spatial distribution. We next measure *accessibility of job openings* for job seekers residing in different locations and using different transportation modes, and explore its patterns of variations. The analyses focus on opportunities and accessibility for *low-skilled people*. We conclude the paper with a discussion of the planning and policy implications of our research findings.

2. The Geography of the Urban Labor Market from the Perspective of Job Seekers

Measuring the quantity and spatial distribution of employment opportunities is an essential element of any serious research on the geography of the urban labor market. There are different ways to count employment opportunities, and the results can be drastically different. Shen (2001) notices that the two most commonly used measures are (1) employment change over a time period and (2) current employment level. He further observes that while researchers are often ambiguous about the reason for choosing one of these two utterly different approaches, there appears to be a strong connection between how job opportunities are quantified and what conclusions about patterns of variations in employment accessibility are drawn. Specifically, researchers who hold the mainstream view about "spatial mismatch," including Hughes (1995) and Kasarda (1995), usually develop their arguments based on employment changes, whereas researchers who challenge the mainstream view, including Shen (1998) and Turner (1998), often build their cases based on employment levels.[2]

[2] Because these researchers focused on barriers to work participation of low-skilled people, they included only low-skilled jobs in the calculation of employment changes or employment levels.

Because new jobs are much more spatially dispersed than pre-existing jobs, it is not surprising that the first group of researchers find the central city to be a disadvantaged residential location with respect to employment accessibility, whereas the second group of researchers find the opposite to be true. The question, then, is which of these two measures—employment change or employment level—should be used. Shen (2001) argues that the answer to this question depends on the research objective. If the objective is to examine employment accessibility for low-skilled people in general, only employment level will appropriately measure job opportunities. On the other hand, if the objective is to understand specifically the situations of those low-skilled people who are seeking jobs, neither employment level, nor employment change, will appropriately measure job opportunities. In this case, the relevant opportunities are job openings, i.e. positions that are currently available to job seekers.

Recognizing unemployed workers as the primary target group of antipoverty policy, Shen (2001) proposes to study the geography of the urban labor market from the perspective of job seekers. He develops an analytical framework for estimating job openings, and applies it to a study of employment opportunities and accessibility for low-skilled job seekers in the Boston metropolitan area. In his analytical framework, job openings consist of opportunities created by growth and opportunities created by turnover. He finds that growth, in comparison with turnover, is only a minor source of job openings. In addition, he finds that because turnover is the dominant contributor to job openings, the spatial distribution of job openings primarily reflects the distribution of pre-existing employment, which is relatively concentrated in the central city. Furthermore, he indicates that the central city still gives its low-skilled residents some location advantage with respect to accessibility of job openings. However, he concludes that the primary determinant of low-skilled job seekers' position in the geography of the urban labor market is transportation mobility, not the residential location. For those who depend on public transportation, which typically has slow and infrequent service, the level of employment accessibility is usually very low even if they live in a relatively advantaged residential location.

A potential issue concerning Shen's findings is that his analyses were based on the single case of the Boston metropolitan area. The extent to which the Boston case is different from other U.S. metropolitan areas is unclear. Given the critical importance of the job seekers' perspective on the geography of the urban labor market and the significant policy implications of the resulting new insights, it is imperative to carefully examine other metropolitan areas.

In this research, we extend Shen's work by studying another metropolitan area and comparing the results with those reported in his paper.

3. Analytical Approach

We asked essentially the same questions as Shen did in his research on the Boston metropolitan area, and applied his analytical framework to our study. Although the analytical framework has been fully described in one of his recent publications (Shen 2001), we repeat part of the description here for the reader's convenience.

3.1 Estimation of the number of job openings

Job openings (Oi(t)) in geographic location i at time t can be expressed by the following equation:[3]

$$O_{i(t)} = O_{i(t)}^{growth} + O_{i(t)}^{turnover} \quad (1)$$

where

$O_{i(t)}^{growth}$ is the number of job openings that come from net employment growth, and

$O_{i(t)}^{turnover}$ is the number of job openings that come from turnover, in geographic location *i* at time *t*.

In order to quantify these two components, it is essential to estimate the average rates of employment growth and turnover, as well as the average duration of job vacancies. Using the month as the unit of measurement for time, the product of the monthly employment growth and the vacancy duration yields the value of the first component of the total number of job openings. The product of the monthly turnover and the vacancy duration yields the value of the second component of the total number of job openings.

Under normal macroeconomic conditions, average vacancy duration in the U.S. is roughly 0.5 month, or 15 days.[4] Assuming that the employment level increases or decreases by a constant amount every month during a given time period, job openings due to employment growth can be estimated as follows:

$$O_{i(t)}^{growth} = \frac{E_{i(t)} - E_{i(t')}}{(t - t') \text{ X } 12 \text{ months}} \text{ X } 0.5 \text{ month} \quad (2)$$

[3] It is important to note that employment growth can have negative values (i.e., employment decline and/or relocation), which means that $O_{i(t)}^{growth}$ can be negative.

[4] An empirical study by Abraham (1983) indicates that half a month is a reasonable estimate of the average vacancy duration in the United States. To be sure, vacancy duration can vary substantially among job categories, regions, and time periods. Everything else being equal, a longer vacancy duration means that a larger pool of job openings accumulates. However, vacancy duration does not affect the relative importance of net employment growth and turnover, nor does it influence intrametropolitan distribution of job openings.

where

t is the ending point (year) of the time period,
t' is the starting point (year) of the time period,
$E_{i(t)}$ is employment level in geographic location *i* at the ending time,
$E_{i(t')}$ is employment level in geographic location *i* at the starting time.

While systematically collected data on turnover does not exist, some information is available to help make sound estimates. Until 1981, the U.S. Bureau of Labor Statistics (BLS) conducted annual surveys on turnover in the manufacturing sector. The data indicated that the average monthly turnover rate was roughly four percent, if all components (quits, discharges, and layoffs) were taken into consideration; it was close to three percent if only quits and discharges were taken into consideration. The latter figure is more relevant to this study because quits and discharges lead to job openings, whereas layoffs do not. More recent studies by Anderson and Meyer (1994) and Holzer (1996) suggest that three percent is still a reasonable estimate of monthly turnover rate. Therefore, job openings created by turnover are estimated as follows:

$$O_{i(t)}^{turnover} = 3\% \text{ per month X } E_{i(t)} \text{ X } 0.5 \text{ month} \qquad (3)$$

3.2 Measurement of Job Accessibility

The demand-adjusted measure of accessibility is especially suitable for indicating job seekers' positions in the geography of the urban labor market, because it takes into consideration competition for limited opportunities (Shen 1998, 2001). The following two equations are used to calculate accessibility for job seekers who are, respectively, automobile drivers and captive public transit riders:

$$A_i^{auto}=\sum_j \frac{O_{j(t)} \text{ X } f(C_{ij}^{auto})}{\sum_k [a_k P_{k(t)} \text{ X } f(C_{kj}^{auto})+(1-a_k)P_{k(t)} \text{ X } f(C_{kj}^{tran})]} \qquad (4)$$

$$A_i^{tran}=\sum_j \frac{O_{j(t)} \text{ X } f(C_{ij}^{tran})}{\sum_k [a_k P_{k(t)} \text{ X } f(C_{kj}^{auto})+(1-a_k)P_{k(t)} \text{ X } f(C_{kj}^{tran})]} \qquad (5)$$

where

A_i^{auto} and A_i^{tran} are accessibility scores for job seekers who are automobile drivers and captive public transit riders, respectively, living in location *i*; *i* = 1, 2, ..., N;
$O_{j(t)}$ is the number of job openings in location *j* at time *t*; *j* = 1, 2, ..., N;
$f(C_{ij}^{auto})$ and $f(C_{ij}^{tran})$ are impedance functions for automobile drivers and public transit riders, respectively, traveling between i and j;

a_k is the percentage of households in location *k* that own at least one motor vehicle;

$P_{k(t}$ is the number of job seekers living in location *k* at time *t*; *k* = 1, 2, …, N;

$f(C_{kj}^{auto})$ and $f(C_{kj}^{tran})$ are impedance functions for automobile drivers and public transit riders, respectively, traveling between *k* and *j*.

Equations (4) and (5) together reflect essentially each job seeker's proximity to the job openings for which he or she is qualified, relative to the number of job seekers competing for these positions. Proximity is measured by required travel time, which is determined jointly by travel distance and travel mode. Shen (1998) shows that the expected value of accessibility scores calculated using these equations equals the ratio of the total number of opportunities to the total number of opportunity seekers in the whole geographic area. For this study, the expected value is the ratio of the sum of job openings to the sum of job seekers (A = O/P, where $O = \sum_j O_{j(t)}$ and $P = \sum_k P_{k(t)}$). The ratio provides a benchmark for examining how the accessibility of job openings varies among residential locations and between travel modes.

The number of job seekers living in each location is measured using the number of unemployed workers living in that location. Because the focus of this study is on economic opportunities and accessibility for low-skilled job seekers, it is essential to estimate the number of unemployed workers in each location who are competing for positions that require relatively little formal training. For this study, low-skilled job seekers are defined in operational terms as unemployed workers who seek positions in sales, service, and labor-intensive occupations.[5] The number is quantified for each location on the basis of the assumption that the occupational distribution of unemployed residents resembles the occupational distribution of employed residents.

It is also essential to estimate the number of job openings in each location that are suitable for low-skilled job seekers. As an approximation, the number equals the sum of currently available positions in sales, services, and labor-intensive occupations.

Note that in the remainder of this paper, for the sake of simplicity, the term job seekers will denote low-skilled job seekers, and the term job openings will denote job openings suitable for low-skilled job seekers, unless otherwise noted.

The level of automobile ownership for job seekers is estimated by exploring two alternative approaches. The first approach is to assume that job seekers living in each location have the same level of automobile ownership as the rest of the residents. The other approach is to assume that all job seekers have the same level of automobile ownership (for example, 50 percent) regardless of where they live.

This study uses the travel time threshold function to specify the travel impedance. With a threshold travel time of C, the value of $f(C_{ij}^{auto})$ is 1 when C_{ij}^{auto} is less

[5] The Standard Occupational Codes (SOC) for these occupations are SOC 243-302 (sales), SOC 403-472 (services), SOC 473-502 (farming, forestry and fishing), and SOC 703-902 (construction and machine operation).

than C; it is 0 otherwise. The value of $f(C_{ij}^{tran})$ is similarly assigned. To determine whether the results are sensitive to the definition of the threshold time, three alternative values of C—15, 30, and 45 minutes—are used.

3.3 Visualization of Job Openings and Accessibility

Geographic Information Systems (GIS) provide an effective visualization tool for understanding the spatial distribution of job openings and for exploring patterns of variations in job accessibility. Three sets of maps will be generated. The first set depicts the spatial distributions of job openings created by employment growth, job openings created by turnover, and the sum of these two components. The second set shows the spatial distribution of job seekers and the locations of opportunity-rich and opportunity-poor areas. The final set displays patterns of variations in accessibility of job openings.

Because the spatial distributions of job openings and job seekers are highly skewed, we use three-dimensional (3-D) representations for the first two sets of maps. In the 3-D density maps, the densities of job openings and job seekers in each location (zone) are depicted by the heights, whereas the numbers of job openings and job seekers in each location (zone) are displayed by the volumes, each of which is a multiplication of height and area.

3.4 The Case Study, the Data, and Computation

Ideally, we would like to examine several U.S. metropolitan areas whose geographies have distinctive characteristics, and draw some general conclusions based on a comparison of the results for these different cases. Unfortunately, due to time and resource constraints, we were able to study only a single case. In fact, even the choice of this single case was dictated to a large extent by data availability rather than by the amount of additional insight it can potentially generate. We examined the nine-county San Francisco Bay Area, which is quite different from the Boston metropolitan area in several important aspects and hence can be a good case for comparison. We would probably have chosen a case more strikingly different from the Boston metropolitan area, such as Detroit, Los Angeles, or Phoenix, if data were readily available for them.

The San Francisco Bay Area is a large, multi-centric metropolis with several major cities in addition to the central city of San Francisco. It has a land area of approximately 7,000 square miles and a population of almost 7 million people. In comparison with the Boston metropolitan area, the Bay Area not only has more land, more people, and more large cities, but also has had a faster pace of employment growth. Since 1990 was the most recent year for which all the required data are available, we used 1990 and 1980 data to examine job openings and accessibility for job seekers in the Bay Area. Data on employed workers by occupation, unemployment, and automobile ownership—all by residential location—for 1990 were extracted from the Census Transportation Planning Package

(CTPP). Data on employed workers by occupation by work location for 1990 and 1980 came from, respectively, the CTPP and the Urban Transportation Planning Package (UTPP). In the 1990 data, the Bay Area is divided into 1,099 regional travel analysis zones (RTAZs). Zone-to-zone travel time matrices for automobile and public transit were obtained from the Metropolitan Transportation Commission (MTC), which is the regional transportation planning agency for the Bay Area.

Data conversion was required because the geographic unit for 1980 UTPP data is census tract and the geographic unit for 1990 CTPP data is RTAZ. Using equivalence tables provided by MTC, the 1980 census tract data were allocated to the 1990 RTAZs. Due to missing values in the 1980 UTPP and 1990 CTPP data, 43 out of 1,099 RTAZs were excluded from the calculations of job openings and accessibility.[6] These missing areas are largely located at the fringe areas of Marin, Napa, Sonoma, and Solano Counties.

Accessibility values were calculated using C programs, and maps were generated using the GIS software ArcView.

4. Findings

4.1 The Composition of Job Openings

Between 1980 and 1990, the Bay Area added roughly 650,000 jobs. Estimated using Equation (2), on a typical day in 1990 the number of job openings due to employment growth was roughly 2,720. Among these job openings, 790 positions, or 29 percent of the total, were in sales, services, and labor-intensive occupations. These job openings are considered as suitable for low-skilled workers.

The estimated number of job openings created by turnover, based on Equation (3), was much greater. On a typical day in 1990, there were approximately 44,930 job openings resulting from quits and discharges. About one third of them, roughly 15,430 positions, were suitable for low-skilled workers.

The results are displayed in Table 1. All together, the estimated total number of job openings on a typical day in 1990 was approximately 47,650. Almost one third of them, roughly 16,220 positions, were suitable for low-skilled job seekers.

These figures reveal the striking difference between the number of job openings created by growth and the number of job openings generated by turnover. Employment growth accounted for only 6 percent of total job openings and 5 percent of openings in sales, services, and labor-intensive occupations. These percentages are only slightly higher than the corresponding p ercentages for the Boston

[6] Therefore, 1,056 of 1,099 zones are included in the analyses. The excluded RTAZs are zones 7, 157, 252, 320, 329, 528, 676, 787-789, 793, 795, 864, 868, 902, 907, 918, 955, 963, 969, 985-996, 1004, 1018, 1043-1048, 1068, and 1087, which are mostly located at the periphery of the Bay Area.

Table 1. Estimated job openings in the San Francisco Bay Area on a typical day in 1990.

	Position created by growth	Vacancies created by turnover	Total number of job openings
Total	2,720	44,930	47,650
Suitable for low-skilled	790	15,430	16,220

Note: These results are based on data for 1056 of the 1099 RTAZs; 43 zones are excluded due to missing values in the 1980 UTPP and 1990 CTPP data.

metropolitan area (Shen 2001). The results indicate that, even for a fast-growing metropolitan area like the Bay Area, employment growth was only a minor source of job openings. The great majority of available employment opportunities for job seekers were created by turnover.

4.2 Spatial Distributions of Job Openings and Job Seekers

Table 2 summarizes the intrametropolitan distributions of low-skilled job seekers and job openings suitable for them. It indicates that job openings created by employment growth were much more spatially dispersed than job openings created by turnover. While 13 percent of new opportunities were located in San Francisco, 17 percent of vacancies resulting from turnover were found in the central city. In addition, it shows that approximately 17 percent of all available opportunities were located in the central city, reflecting the fact that turnover was the dominant contributor of job openings.

Furthermore, Table 2 shows that low-skilled job seekers and the job opportunities suitable for them were almost equally concentrated in the central city. Approximately 16 percent of low-skilled job seekers were located in San Francisco, in comparison with 17 percent of job openings.

In comparison with the Boston case, job openings and job seekers located in the central city of San Francisco accounted for somewhat lower percentages of the totals for the whole metropolitan area.[7] This difference can be explained by the relative importance of other major cities, especially San Jose and Oakland, in the Bay Area. However, the two cases show an important similarity: for the central city, the percentage share of job openings was almost the same as the percentage share of job seekers. More detailed pictures of spatial distribution of job openings suitable for low-skilled job seekers are visualized in three maps. These three maps use the same data classification and 3-D visual presentation and, therefore, are directly comparable to one another. Figure 1 illustrates the density and volume of new positions created by employment growth. Figure 2, on the other hand, presents the density and volume of job openings due to turnover. It is apparent that

[7] Approximately 20 percent of job openings for low-skilled workers and 21 percent of low-skilled job seekers in the Boston metropolitan area were located in the central city (Shen 2001).

Table 2. Intrametropolitan distribution of low-skilled job seekers and job openings in the San Francisco Bay Area on a typical day in 1990.

	Entire Bay Area	Within San Francisco	Outside San Francisco
Position created by growth	790	110 (13%)	680 (87%)
Vacancies created by turnover	15,430	2,680 (17%)	12,750 (83%)
Job openings for the low-skilled	16,220	2,790 (17%)	13,430 (83%)
Low-skilled job seekers	64,720	10,540 (16%)	54,180 (84%)

Note: These results are based on data for 1056 of the 1099 RTAZs; 43 zones are excluded due to missing values in the 1980 UTPP data.

employment growth was a relatively minor source of job openings, as discussed earlier. It is also apparent that growth was more spatially dispersed than turnover. Job openings created by turnover were highly concentrated in the municipalities surrounding the bay, with the highest densities found in downtown San Francisco and downtown Oakland. Figure 3 displays the density and volume of job openings in total. Because of the dominance of turnover as a source of available opportunities, Figure 3 looks quite similar to Figure 2. These results are generally consistent with those obtained from the case of the Boston metropolitan area (Shen 2001).

The spatial distribution of low-skilled job seekers is visualized in Figure 4. Not surprisingly, low-skilled job seekers were highly concentrated in the three largest cities in the Bay Area, with the highest densities found in San Francisco. What might be surprising, however, is that many other municipalities, including Santa Rosa, Vacaville, and Pittsburg, that are located far from the central city, also had considerable numbers of low-skilled job seekers. In fact, virtually all suburban and periphery zones showed at least a modest density of job seekers (20–40 persons/square mile). These results are also broadly consistent with those obtained from the case of the Boston metropolitan Area (Shen 2001).

Figure 5 portrays the ratio of job openings to job seekers in each zone. The average ratio was approximately 0.25 for the whole Bay Area with an estimated 16,220 available positions in low-skilled occupations and 64,720 seekers of these positions. The map shows that most of the zones with relatively high ratios of opportunities to seekers, including downtown San Francisco and South San Francisco, were located near the Bay. There were noticeable exceptions, including Livermore and Napa, which were quite far from both the central city and the Bay. In comparison with the Boston metropolitan area, the Bay Area has more scattered zones with high job openings to job seekers ratios.

Figure 1.

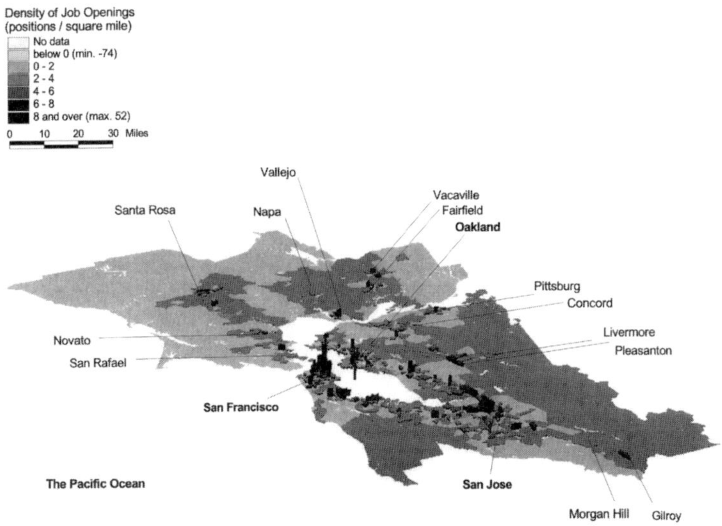

Figure 1. Density of job openings due to employment change

Figure 2.

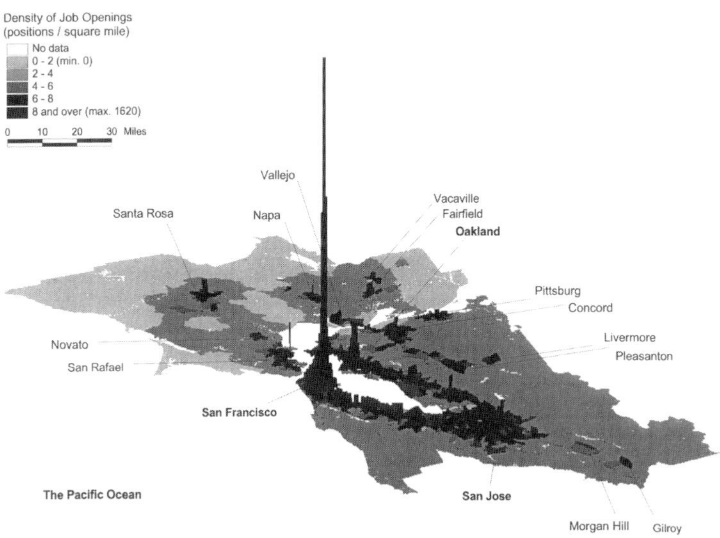

Figure 2. Density of job openings due to turnover

Figure 3.

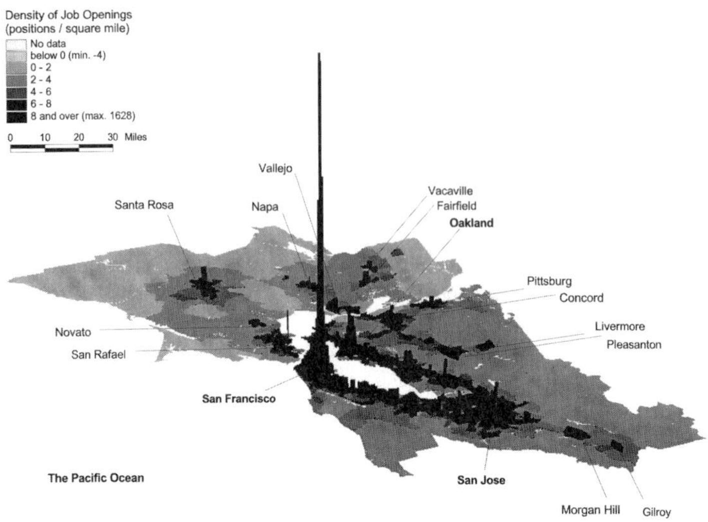

Figure 3. Density of job openings in total

Figure 4.

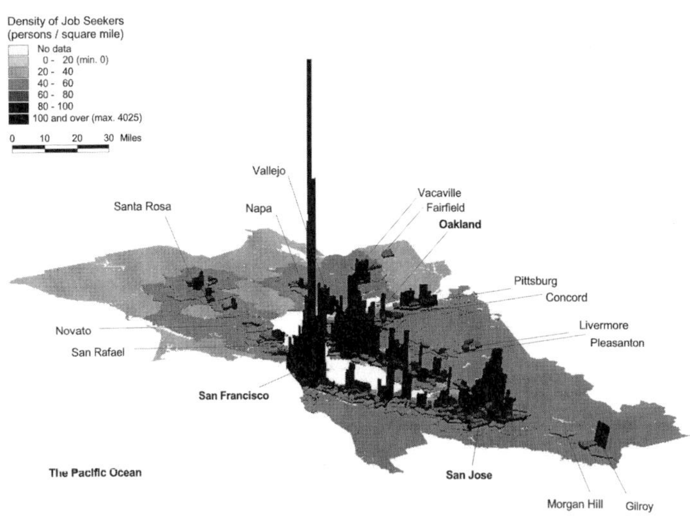

Figure 4. Density of job seekers

Figure 5.

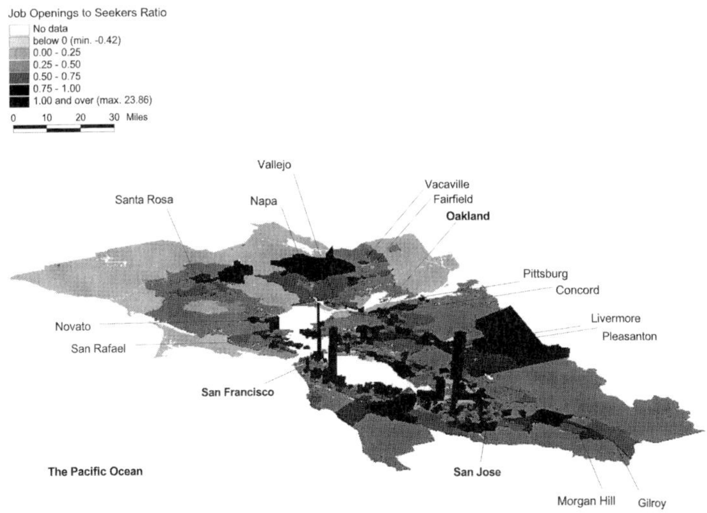

Figure 5. Ratio of job openings to job seekers

From Figure 5, it is also apparent that low-income neighborhoods in the city of San Francisco are located in proximity to high concentrations of job openings in downtown San Francisco and South San Francisco. However, low-income neighborhoods in some other parts of the Bay Area, including those in south Oakland, are not in geographic proximity to high concentrations of employment opportunities.

4.3 Spatial Variations in Accessibility of Job Openings

It is well know to urban geographers and urban transportation planners that geographic distance is just one of the factors that determine accessibility. Transportation mobility and competition for the available opportunities are two other key factors. An opportunity seeker's position in the geography of the urban labor market is determined jointly by these variables, and is measured by either Equation (4) or Equation (5), depending on the seeker's transportation mode.

Figure 6 portrays spatial variations in employment accessibility for job seekers who traveled by car, and Figure 7 portrays spatial variations in employment accessibility for job seekers who depend on public transportation.[8] Individually, each map shows that, for job seekers who traveled by the same transportation mode, accessibility of job openings generally increases as the residential location gets closer to the central city. In other words, low-skilled job seekers would likely worsen rather than improve their positions in the geography of the urban labor market by relocating to suburban or periphery areas.

Together, these two maps indicate clearly that accessibility differentials among workers residing in different locations were relatively minor in comparison with accessibility differentials among workers traveling by different transportation modes. The average value of accessibility for job seekers who traveled by car was 0.28, whereas the average for job seekers who traveled by public transit was merely 0.02. In fact, 588 out of 1099 RTAZs were "accessibility-rich zones" for auto drivers, i.e. accessibility for job seekers who lived in these zones and used private vehicles for commuting was higher than the overall average of 0.25. On the other hand, none of the RTAZs was accessibility-rich for public transit riders. These results are also similar to what Shen (2001) found in his study of the Boston metropolitan area. They provide new evidence in support of his two observations. First, residing in central cities still gives low-skilled job seekers some limited location advantage in access to employment opportunities. Second, job seekers who depend on public transit are spatially disadvantaged almost no matter where they live, because transportation mobility is the key determinant of people's positions in the geography of labor market in a dispersed metropolitan economy.

[8] The results shown in these two maps were based on the assumption that the threshold travel time was 30 minutes. A threshold travel time of 15 minutes or 45 minutes resulted in somewhat different maps but basically the same conclusions. It is also important to point out that, because 59 zones were excluded in the calculation of accessibility due to missing values, the maps display slightly incomplete and perhaps even slightly distorted pictures of patterns of variables in accessibility for low-skilled job seekers. The distortion is likely to be small, however, because most of the excluded zones are located in periphery areas that are more than 30 minutes away from other zones. These excluded zones thus have little effect on the calculation of accessibility for the remaining zones.

Figure 6.

Figure 6. Job accessibility for low-skilled job seekers who use automobiles for commuting

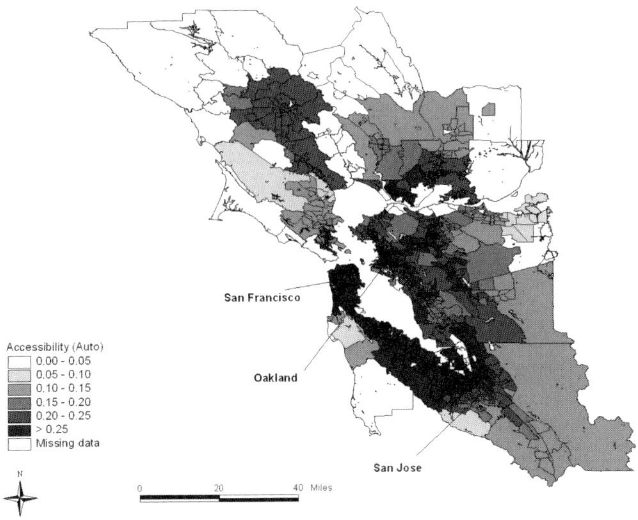

Figure 7.

Figure 7. Job accessibility for low-skilled job seekers who use transit for commuting

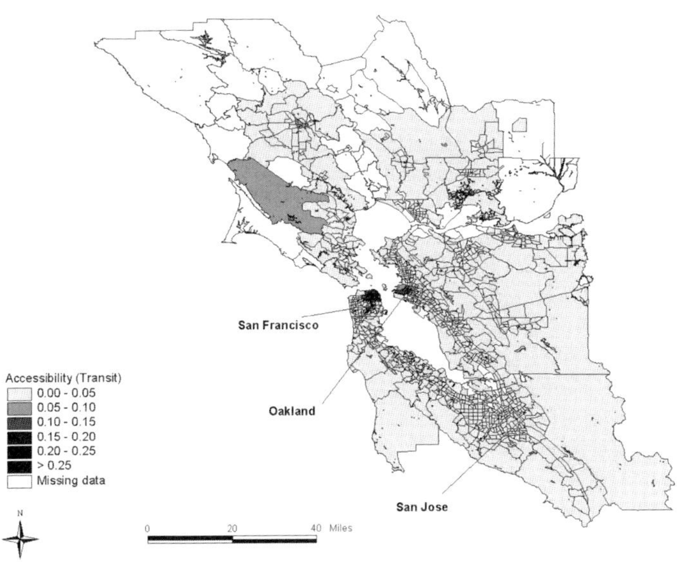

Conclusions

This research has strengthened the empirical basis for challenging some of the popular views on the geography of the urban labor market that have been highly influential in both academic and policy arenas. The findings from the case study of the San Francisco Bay Area are generally consistent with the findings obtained from a recent study of the Boston metropolitan area (Shen 2001). They demonstrate that the great majority of currently available employment opportunities—job openings—come from turnover. Specifically, estimation of the number of job openings in the Bay Area on a typical day in 1990 shows that turnover accounted for approximately 95 percent of available employment opportunities for low-skilled job seekers. In comparison, employment growth is only a minor source of job openings.

In addition, the research findings indicate that, because turnover is the dominant source of job openings, the spatial distribution of available opportunities primarily reflects the spatial distribution of preexisting employment. Therefore, the available opportunities are still relatively concentrated in the nonresidential parts of central cities. While low-skilled job seekers are also highly concentrated in central cities, especially in the low-income residential neighborhoods within which there are few jobs, many of them live near some opportunity-rich areas, such as downtown San Francisco. If they are willing and able to travel beyond their neighborhood boundary, they can tap into a much larger pool of potential jobs. On balance, assuming a normal travel-time budget, low-skilled job seekers living in central cities have some location advantage with respect to accessibility of job openings in comparison with their suburban counterparts. This is especially true for job seekers who do not have a car, because bus service is generally much better in the metropolitan center than at the periphery. However, there are significant variations in accessibility among central-city neighborhoods as well as among peripheral locations.

Furthermore, the findings clearly show that accessibility differentials among locations within a metropolitan area are rather small when compared to accessibility differentials between travel modes. The data for the Bay Area demonstrate once again that for job seekers who travel by car, most residential locations will allow them to have an above-average level of accessibility. For job seekers who depend on public transit, on the other hand, accessibility is substantially below the average almost regardless where they live.

These findings have important methodological and policy implications. First of all, it is incorrect to use employment growth to measure the overall quantity and spatial distribution of currently available opportunities in a metropolitan labor market. Not only does employment growth account for merely 5 percent of job openings, but it also presents an overly optimistic picture of economic opportunities in the suburbs and an overly pessimistic picture of economic opportunities in central cities.

Second, in any empirical measurement of workers' positions in the geography of the urban labor market, it is essential to differentiate those who travel by car from those who depend on public transit. A mixture of these very different groups, as it has often been the case in many studies, will inevitably lead to underestimation of the spatial barrier facing the spatially most disadvantaged group—those who do not have private automobiles as an option for travel. It may also lead to analyses that mistakenly attribute low-income people's spatial disadvantage to their residential location, even though the true cause is the gap in transportation mobility between private automobiles and public transit.

Third, in examining the geography of the urban labor market, the data should be processed at an adequate level of spatial resolution to capture differences among central-city neighborhoods and among peripheral locations. The researcher should also be aware of the fact that there are important differences among job seekers in terms of family responsibility and hence the allocation of travel-time budget. A normal length of commute time may not apply to a single parent who needs to send several children to school and daycare center before going to work.

Fourth, perhaps more fundamentally, researchers should realize that there is an increasingly important social dimension in urban spatial analysis and modeling in general and in transportation research in particular (Shen 2000). It is highly likely that population groups with different transportation mobility characteristics will exhibit diverging location preferences in the future. If urban spatial analysis and modeling is to serve as an effective tool for planning and policymaking, it must be capable of helping planners and policymakers understand the consequences of these diverging location preferences.

Fifth and finally, mobility-enhancement policies and programs must clearly target the public-transit-dependent people. The objective is to narrow the gap in accessibility between the target group and the rest of the population. Conceptually, this can be achieved by taking several alternative approaches, which include lowering the barriers to automobile ownership, improving public transportation service, encouraging ride sharing, and integrating public transportation service with housing development. There is an urgent need for serious research to help choose among these approaches. Our study has shown that, even though job opportunities and job seekers are still relatively concentrated in central cities, they are nevertheless spatially distributed in the whole metropolitan area. This makes it impossible to bridge the accessibility gap effectively by adding just a few bus or vanpool services. Will some other alternatives work better? Can they work together? These are some of the questions that need to be answered in future research.

Acknowledgements

We wish to thank Chuck Purvis of the Metropolitan Transportation Commission (MTC) for the San Francisco Bay Area for providing part of the data used in this research. We also wish to thank Jee-Seong Chung, a graduate student in the Department of Urban Studies and Planning at MIT, for his able assistance in mapping.

References

Abraham, K. G. 1983. Structural/frictional vs. deficient demand unemployment some new evidence. *The American Economic Review* 73:708–724.

Anderson, P. and B. Meyer. 1994. The extent and consequences of job turnover. *Brookings papers on economic activity—Microeconomics*. 177–248.

Batty, M. 1976. *Urban modelling*. Cambridge, UK: Cambridge University Press.

Blumenberg, E. and P. Ong. 1998. Job accessibility and welfare usage: Evidence from Los Angeles. *The Journal of Policy Analysis and Management* 17:639–657.

Holzer, H. J. 1996. *What employers want? Job prospects for less-educated workers*. New York: Russell Sage Foundation.

Hughes, M. A. 1995. A mobility strategy for improving opportunity. *Housing Policy Debate* 6:271–297.

Kasarda, J. 1995. Industrial restructuring and changing location of jobs. *State of the Union: America in the 1990s*, Volume I: Economic Trends. Edited by R. Farley. New York: Russell Sage Foundation.

Lacombe, A. 1997. *Welfare reform and access to jobs in Boston*. Washington, D.C.: Bureau of Transportation Statistics, U.S. Department of Transportation.

Landis, J. D. 1994. The California urban futures model: A new generation of metropolitan simulation models. *Environment and Planning B* 21:399–420.

Lowry, I. S. 1964. *A model of metropolis*. Santa Monica, CA: Rand Corporation.

Putman, S. 1983–1991. *Integrated urban models*. London: Pion.

Rich, M. J. and J. Coughlin. 1998. The spatial distribution of economic opportunities: Access and accessibility issues for welfare households in metropolitan Atlanta. Paper presented at the 94th annual meeting of the Association of American Geographers.

Shen, Q. 2001. A spatial analysis of job openings and access in a U.S. metropolitan area. *Journal of the American Planning Association* 67:53–68.

Shen, Q. 2000. Spatial and social dimensions of commuting. *Journal of the American Planning Association* 66:6882.

Shen, Q. 1998. Location characteristics of inner-city neighborhoods and employment accessibility of low-wage workers. *Environment and Planning B* 25:345–365.

Turner, M. A. 1998. *Do neighborhoods matter? Place, race and economic opportunities in the Washington metropolitan area*. The 1998 Lefrak Monograph, University of Maryland, College Park.

Wachs, M. and B. Taylor. 1998. Can transportation strategies help meet the welfare challenge? *Journal of American Planning Association* 64:15–19.

Wegener, M. 1994. Operational urban models: State of the art. *Journal of the American Planning Association* 60:17–29.

Wilson, A. G. 1974. *Urban and regional models in geography and planning*. New York: John Wiley

Modeling Opportunity: Employment Accessibility and the Economic Performance of Metropolitan Phoenix Neighborhoods

C. Scott Smith

1. Introduction

The concept of accessibility, defined as the ease with which activities can be reached from a specific location, is widely applied by researchers to perform a variety of tasks (USDOT 1997). Commercial planners identify optimal locations for businesses based on the accessibility of target markets. Land-use planners propose locations for public facilities based on user accessibility. Similarly, equity planners utilize measures of accessibility to assess whether resources are fairly distributed throughout an urban area (Cervero, Rood, and Appleyard 1995). In these ways accessibility measures influence land-use decisions that have the potential to empower communities or create vast barriers of social inequity (Beatley 1994).

This research focuses on one particular aspect of accessibility—that is, the relationship between job accessibility and the economic performance of central city neighborhoods. Researchers from a range of disciplines have argued that the job accessibility of central city workers has declined over time due to a host of factors including exclusionary housing practices, limited mobility, and employment suburbanization (Ihlanfeldt 1992; Ihlanfeldt and Young 1999; Kain 1968; Kasarda 1990; Kasarda 1994; Knox 1990; Turner 1997). The ensuing *skills mismatch* is marked by concentrations of low-skill workers living within central cities where high-skill jobs prevail. This is coupled by a *spatial mismatch* of central city workers with inferior access to high-growth suburban areas where there are substantial low-skill job opportunities.

Using metropolitan Phoenix as a case study, this article suggests that job accessibility measures are powerful indicators of economic performance and should therefore be utilized by decision makers in the planning process. More specifically, it argues that indicators of job accessibility can greatly assist decision-makers by providing an assessment of how urban growth may influence the distribution of economic performance and variations in social and environmental equity.

Hypotheses

This paper simultaneously relates measures of job accessibility and economic performance for the purpose of addressing the following two hypotheses:

- Levels of job accessibility decline with respect to occupational skill level such that high-skill workers have greater accessibility to employment than low-skill workers.
- Job accessibility has significant adverse effects on the economic performance of central city workers.

The first hypothesis addresses the job-skills mismatch assumption, which argues that the demand for skills has increased in urban labor markets over time (Burns and Gober 1998; Kasarda 1990; McLafferty 1996; McKenna 1996; Kain 1968). Central city minority residents lag behind other population groups with respect to training and educational attainments and are therefore less likely to procure nearby job opportunities in upper labor market segments (i.e., job opportunities in Phoenix's central business district).

The second hypothesis relates to the spatial mismatch assumption, which argues that high unemployment is correlated with the spatial distribution of employment opportunities (Kain 1968). Previous research has suggested that the suburbanization of employment, exclusionary housing practices, and limited mobility have effectively concentrated low-skill minority workers in the central city with sparse access to job opportunities (Hughes 1990; Hughes 1991; Sheppard 1990; Summers 1993; Wacquant 1993). As a result, central city workers are more likely to live in neighborhoods plagued with high levels of joblessness (Arnott 1998; Bauder 1999; Boswell 1997; Darden, Bagaka, and Shun Jie 1997; McLafferty 1996).

2. Laying the Theoretical Groundwork: A Review of Spatial Mismatch and Job Accessibility Literature

The U.S. economy, by many measures, is a burgeoning system. With a per capita gross national income of $34,260 it is one of the wealthiest countries in the world (World Bank 2001). Low inflation and unprecedented economic growth has permeated the 1990s and is projected to continue, albeit at slightly slower rates from the rapid pace of the past several years. National employment growth consistently outpaces population growth and, at four percent, the annual unemployment rate is the lowest it has been since the Nixon era (Current Population Survey 2000).

Despite these impressive indicators, there is one particular subculture of the population that this prosperity eludes. Ironically, this group resides in some of the most job-rich areas in the country—central cities. Over the past 20 years, researchers have referred to this spatially concentrated, ethnically diverse group as the *urban underclass* (Kasarda 1990; Massey 1993), the *truly disadvantaged* (Wilson 1987), the *underemployed* and the *new urban poor* (Wilson 1996; Knox 1990).

For this group, the American economy imparts a much harsher reality. Predominantly minorities, the central city poor are both economically and spatially excluded from the economic, social, and political mainstream of society. Extreme levels of poverty and joblessness often pervade their residential communities. Many households are headed by single women and a high percentage of the population either receives public assistance or has withdrawn from the labor force (Marcuse 1997). Different from the immigrant and ethnic enclaves that preceded it, this new *outcast ghetto* is socially destructive (Marcuse 1997). Opportunities for a viable education or cultural life are often eroded.

Spatial mismatch: A structural explanation for joblessness

Throughout the political continuum researchers heatedly debate the root causes and continued emergence of today's central city poor. In 1966, the controversial anthropologist Oscar Lewis emphasized that there were personal traits—such as immaturity and unreliability—associated with cultures of poverty (Lewis 1966). Conservatives adopted this hypothesis, claiming that efforts to eliminate poverty through social policy are unproductive. Charles Murray's *Losing Ground*—which became a conservative icon during the Reagan administration—suggests that social programs of the 1960s and 1970s fostered dependence and welfarism and thus did little to improve the condition of poorer populations (Murray 1984). This and related supply-side or individual-based explanations for minority joblessness continued to find support in the 1990s and culminated in the Personal Responsibility Act passed by the 104th Congress and signed by President Clinton in 1996.

In contrast, liberal research has primarily focused on demand-side explanations for central city joblessness, including racial discrimination and spatial mismatch. A Chicago-based study by Ellwood found racial composition to be the dominant explanatory variable for employment rates (1986). Later studies have also evidenced that unemployment is highly correlated with the employment status of working minorities (Bendick, Jackson, and Reinoso 1994; Cervero, Rood, and Appleyard 1995; Neckerman and Kirshenman 1991; Turner 1997).

However, most theorists place greater emphasis on the skills and residential locations of workers than on discriminatory practices and attitudes of employers (Meiklejohn 2000). Spatial mismatch, for instance, is the dominant demand-side explanation for the prevalent joblessness found in many central city neighborhoods. According to this hypothesis, central city residents, with lower educational attainments and skill sets, typically do not have the professional requirements to procure jobs in the upper labor markets that tend to dominate most central business districts. Furthermore, housing segregation coupled with the limited mobility of minority residents—marked by low vehicle ownership and inadequate public transit systems—inhibits their ability to procure low- to moderate-skill jobs that have progressively migrated outside the central city.

Kain first posited the hypothesis that central city unemployment is correlated with job accessibility (Kain 1968). Using Detroit and Chicago as case studies, Kain found that the negative effect of housing segregation on African American employment was magnified by the decentralization of jobs. The hypothesis gained momentum in the mid to late-1980s and was assigned the moniker *spatial mismatch*.

Since its inception over 30 years ago, there have been more than 50 published spatial mismatch studies. The methodogies used in these studies reflect the popular epistemologies and analysis tools of the time. A thorough examination and classification of these methods was undertaken by Ihlanfeldt (1992) and Stoll (1999). Together, these studies have yielded inconclusive results. However, Ihlanfeldt argues that much of the research made inappropriate assumptions and suffered from flawed methodologies (1992). If the flawed studies are dismissed, Ihlanfeldt claims that the empirical evidence lends credibility to the spatial mismatch hypothesis.

Advances in measuring job accessibility

Job accessibility is defined as the ease with which employment opportunities can be reached. It is regularly identified as an important element of smart growth and sustainable development principles. Roseland, for example, suggests that cities should work to increase job accessibility in order to improve air quality and worker productivity by relieving traffic congestion and reducing commute times (1998). Transit-oriented development, mixed-use zoning, and alternative-mode provisions are strategies that planners are promoting to better integrate land use and transportation, thereby enhancing worker access to employment opportunities.

While the benefits of advancing job accessibility are generally understood, researchers are less likely to reach consensus on the appropriate methods to measure it.[1] Instead, researchers implement a wide variety of approaches that vary with respect to geographic and demographic complexity. A large share of these approaches can be separated into two basic classes: direct and indirect measures. Direct accessibility calculations—the kind applied in this study—report variations of employment availability and demand on an interval-scale by *neighborhood*. Neighborhoods are commonly represented as census tracts or traffic analysis zones (TAZs). Because direct measures are interval-scale and are reported at relatively detailed levels of geography, they allow planners to visually assess a continuous variation of job accessibility throughout an urban environment.

In contrast, indirect measures of accessibility generalize space by grouping residential locations of workers into broad categories such as "central city" or

[1] For a thorough review of these approaches refer to Ihlanfeldt (1992).

"suburbs." This approach does not account for the physical transportation system or the actual costs of commuting to specific employment opportunities. Indirect measures also fail to consider the level of competition between similarly skilled workers and thus potentially underestimate job demand. Direct measures, therefore, are more likely to reflect the urban environment in that neighborhoods are modeled with respect to their geographic and demographic context in the metropolitan area.

To better understand the versatility and limitations of direct job accessibility measures, however, it is important to consider their characteristic flaws. Direct measures have been criticized in previous research for the three following reasons. First, these measures typically do not account for the endogeneity of residential location in the employment process. This so-called simultaneity effect suggests that people may opt to reside in communities with inferior job accessibility in order to consume more affordable housing and other reasons, which consequently muddles the effects of job access on economic performance. If this factor is ignored, effects of job access on economic performance can be underestimated (Ihlanfeldt 1992; Stoll 1999).

A second criticism of direct measures of accessibility concerns the assumption that the demand for employment opportunities is uniformly distributed and that job opportunities have no capacity limitation (Shen 1998). Workers, however, are not evenly distributed throughout the urban environment nor do they have similar skill sets. For example, it is probable that low-skill central city workers have poor job accessibility even though they live in job-rich areas (e.g., most central business districts). Additionally, no single job opportunity can support more than one employee. Therefore, neighborhoods, because of their unique locations within the region, will exhibit disparate levels of job demand.

The third criticism relates to the typically auto-centric calculation of direct accessibility. That is, commute times are often confined to automobile commutes and thus fail to consider the greater frictional factors associated with alternative transportation modes such as public transit, walking, or bicycling. An automobile owner, for example, typically has greater access to employment opportunities than his/her neighbor who commutes by bicycle or public transportation.

This study, given its scope and data limitations, does not address all of the known issues outlined above. Given that the simultaneity effect is only expected to enhance the predictive ability of job access, ignoring this factor does not discredit this study's results. Also, unlike previous studies, this research accounts for employment demand, variable skill levels, and multiple modes of transportation using refined GIS and statistical procedures.

3. Setting the Stage: Housing, Transportation, and Employment Trends in Metropolitan Phoenix

Metropolitan Phoenix[2] is an appropriate case for studying employment accessibility primarily because it has developed differently than most eastern and midwestern cities that are disproportionately represented in spatial mismatch research. Unlike these cities, metropolitan Phoenix did not become a large urban center until after World War II, when Arizona became the nation's fastest growing state. With well over 500,000 new jobs created during the 1990s, metropolitan Phoenix ranks among the nation's leaders in population and economic growth.

Also, public transit, save for a period between 1890 and 1940, has led a meager existence in the region. The great majority of transportation improvements over the past 30 years have been dedicated to developing the expansive, yet late-blooming, freeway network. The disparity between freeway and public transit spending explains the exceptionally small per capita transit miles in the area. At seven miles, the per capita rate is exceeded by comparable-sized metropolitan areas such as San Diego (11 miles per capita), Seattle (23 miles per capita), and Denver (20 miles per capita) (Morrison Institute for Public Policy 2000).

Metropolitan Phoenix also differs from other regions highlighted in spatial mismatch literature in that the dominant minority population is Hispanic (or Latino) (over 16 percent). The percentage of population that is African American (approximately three percent)—the ethnic group most often identified with the new urban poor—is well below the 1990 national average of 13 percent (U.S. Census Bureau 1990).[3]

Additionally, Phoenix never had strong low-skill industrial and manufacturing sectors (e.g. steel and automobile) as did other cities commonly researched in spatial mismatch literature. Its isolated location in the Southwest has historically been more conducive to air transportation, which is more efficient for small, high value, and often highly technological goods (Gammage 1999).

Nonetheless, metropolitan Phoenix also shares many characteristics with its eastern and midwestern counterparts. For example, the area has a history of institutional racism, which fostered residential ethnic segregation that is conspicuously present in the Valley today (Luckingham 1994). Metropolitan Phoenix has also experienced, although not as profoundly, the impact of post-Fordist economic restructuring from mainly low-skill agricultural and manufacturing occupations to service occupations.

Phoenix's central city also resembles the nationwide urban patterns of decay, redevelopment, and gentrification. The central city experienced a significant

[2] Metropolitan Phoenix refers to Maricopa County, Arizona, which comprises the city of Phoenix and 23 other incorporated communities.

[3] Hispanic is a cultural category and is used here to refer to any Spanish-speaking race. It is also used interchangeably with the term Latino, as in U.S. Census Bureau reports. Census 2000 figures indicate that the percentages of Hispanic and African American populations in Maricopa County increased to 24.8 and 3.7 respectively.

decline after World War II such that by the late 1960s it had no department stores. Retail jobs and shopping centers had gradually migrated to suburban regional malls. It was not until the late 1980s that the city made a concerted effort to redevelop its urban core. With the help of a strong mayor and citizen support the city has redefined itself as the area's sports, entertainment, and government center (Gammage 1999). Moreover, federal dollars have retrogressively shaped the social and economic landscape of metropolitan Phoenix via federal highway administration grants, public housing, and other federal grants as is the case in other cities (Norquist 1998).

Housing segregation: Central city ghettoes and suburban citadels

One need only to observe distributional maps of ethnicity and income to conclude that there are cogent forms of segregation at work in metropolitan Phoenix. Minority and lower-income residents are concentrated in the urban core while higher-income households reside outside the central city (Figures 1 and 2). In some cases these patterns may be economic and voluntary such as the development of immigrant enclaves due to chain migration and the rational consumption of affordable housing (Marcuse 1997). Alternative explanations for this polarized ethnic and economic landscape include institutional racism, misguided federal housing and transportation policy, and economically exclusive housing practices (Kunstler 1993a; Kunstler 1993b; Schill and Wachter 1995; Norquist 1998).

Luckingham wrote of the evolving residential patterns of Phoenix minorities (Luckingham 1994). In Phoenix's early history, Latino, African, and Chinese Americans were involuntarily forced by the white majority to reside in segregated quarters. The legacy of these quarters, along with the lack of affordable housing and the historic and continued discrimination against minorities by lending institutions (Cuomo 1999), has restricted the residential and economic opportunities of minorities.

For example, educational advancements of central city residents are often restricted because of inadequate schools. A recent report by the Morrison Institute for Public Policy identifies a dramatic achievement gap—evidenced by poor performance on standardized test scores—between white and non-white school districts (2000). The predominantly minority populated areas of central and southwestern metropolitan Phoenix averaged 34 and 35, respectively, on the Stanford 9 Achievement tests. This contrasts the principally white northeast and southeast districts that scored 72 and 59.[4] The report warns:

> The region has to worry about the education of children in central Phoenix and the southwest portion of the region. Individual economic success correlates particularly with educational attainment. The weak schools of the center present

[4] Scores reflect an average of reading and math. A rank of 34 means that on average, students taking the test scored lower than 64 percent of students nationwide.

a powerful impetus for [further] decentralization. (Morrison Institute for Public Policy 2000)

Public transit: Difficulty escaping the urban core

Public transit has had a turbulent history in metropolitan Phoenix. Falling in and out of public ownership and support, the region continues to struggle to find a common ground between providing transit services and freeing traffic for those who drive automobiles (Abbitt 1990). Much of the problem is rooted in the historically less dense pattern of development that is pervasive in the Valley and not conducive to public transportation. In 1985, Maricopa County voters passed legislation to construct 233 miles of freeways. Since then, most state and county money has focused on highway building and neglected transit.

Not only are there fewer transit services in general, but the specific services provided for low-income residents who are less likely to make the capital investment for an automobile are inferior when compared to the services provided for their wealthier counterparts. For instance, several low- to moderate-skill employment opportunities are available outside the central city. However, the region's express buses (30 percent of all bus routes) are targeted toward providing high-skill, higher-income suburban residents an efficient means to commute downtown. These buses do not provide similar services to downtown residents who make suburban commutes. As a result, central city residents have limited access to suburban job opportunities. Transit riders who live in central city and work in the suburbs must often partake in circuitous, multiple bus journeys to their workplaces.

Labor market trends: Considerable growth, considerable distance

The positive national employment trends outlined earlier in this paper are amplified for the state of Arizona. Since the end of the last recession in 1991, the state has experienced a record setting span of continuous economic growth (Center for Business Research 2000). Per capita personal income rose 4.7 percent in the state between 1997 and 1998, which is significantly higher than the 3.6 percent national average. 1999 marked the seventh year of strong expansion indicated by an employment-to-population ratio exceeding 56 percent, a record. Maricopa County's average unemployment rate of 2.9 percent in 1999 compares favorably with the national average of 4.2 percent. Undoubtedly these are impressive numbers. However, the manner in which these trends play out across the urban landscape is a cause for concern.

While many large urban areas are evolving from a monocentric (i.e., one employment core) to a polycentric structure (i.e., multiple cores), Phoenix continues to have a relatively centralized urban form with respect to employment. This may, in part, be due to the fact that the region only recently began constructing freeways. According to a recent study, one third of the region's employment opportunities are located in two central areas of the urban core. The majority of these jobs, however, are in upper labor market segments that require substantial education. Thus:

> There is a potential separation in the region between appropriate job opportunities and the location of less-skilled workers. In metropolitan Phoenix, these less-skilled workers often reside predominantly in or near the downtown Phoenix employment areas. However, the jobs accessible to them are heavily weighted toward professional positions. That raises the possibility of their spatial isolation from needed entry-level work opportunities. (Morrison Institute for Public Policy 2000)

Employment opportunities outside the central city are primarily construction, service, and retail jobs. Although educational attainment for jobs in these sectors are less demanding, the considerable distance of these opportunities from the central city prohibits many downtown residents from procuring them.

4. Recalibrating Employment Accessibility

This paper utilizes direct measures of accessibility that were calculated using two equations. The equations are adapted from Shen's model, which overcomes many of the limitations of direct accessibility measures expressed earlier in the paper (Shen 1998). Unlike Shen's model, however, the following equations consider three levels of mobility, four occupation skill levels, and impedance values that were calculated using rule-based geographic information system (GIS) automobile and transportation networks. Particular details of the models and their divergence from other measures are provided below.

Equation 1 calculates the job accessibility of origin neighborhood i ($A_i^{mode\,t}$) with respect to the percentage of workers with skill set k (W_k) that commute to work using transportation mode t ($a_k^{mode\,t}$). The numerator computes the sum of relevant employment opportunities (E_j) or job availability within the metropolitan area from an origin neighborhood while accounting for impedance costs ($C_{ij}^{mode\,t}$) to procure the opportunities.

$$A_i^{mod\,et} = \sum_j E_j f(C_{ij}^{mod\,et}) / \sum_k (\alpha_k^{mod\,et} W_k f(C_{kj}^{mod\,et}) + (\alpha_k^{mod\,en} W_k f(C_{kj}^{mod\,en})) \quad (1)$$

In contrast, the denominator represents the worker demand for relevant employment opportunities accounting for the sum of impedance costs of workers commuting by transportation mode t and all other analyzed modes of transportation ($C_{kj}^{mode\,t}$). Equation one was repeated for each transportation mode and subset of workers. Given that this study identifies four categories of occupational skill and three categories of transportation mode, the equation yielded 12 sets of accessibility values.

Figure 1.

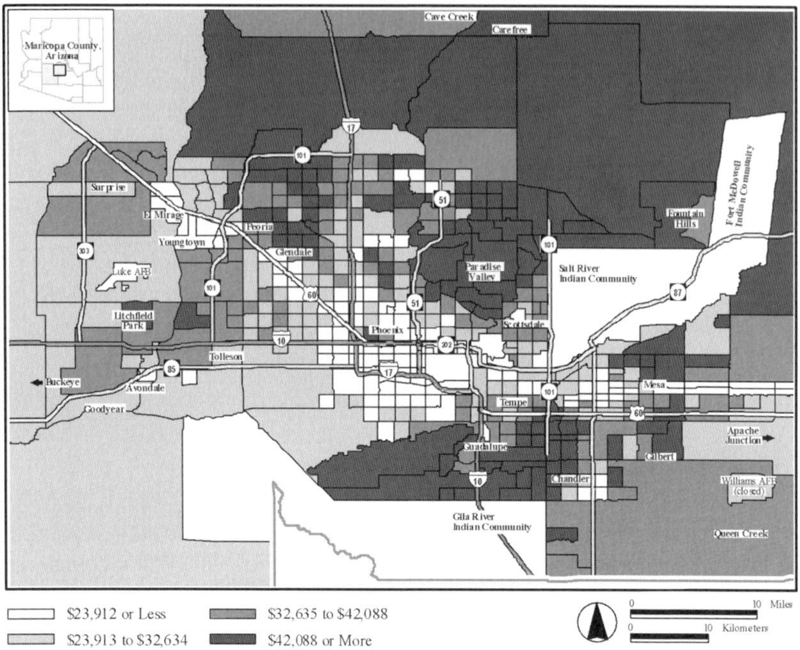

Impedance costs (C) were calculated using an exponential decay function having the form $e^{-\beta d_{ij}}$. Where e is the natural log, β is the exponent for distance decay, and d_{ij} represents the distance between locations i and j. The distance decay exponent changes with transportation mode so that accessibility decreases at a greater rate for low mobility transportation modes, such as bicycling and walking, and at a slower rate for high mobility transportation modes such as the automobile (Figure 1).

Equation 2 calculates the general employment accessibility (A_i^G) of workers with a specific skill set living in neighborhood i while accounting for all transportation modes. It is a composite index that weights results of (1) by the ratio of workers utilizing specific modes of transportation for their work commutes. Therefore, (2) will yield four sets of accessibility values—one for each occupation skill set.

$$A_i^G = (\alpha_i^{\text{mod}\,et} W_i / W_i) A_i^{\text{mod}\,et} + (\alpha_i^{\text{mod}\,en} W_i / W_i) A_i^{\text{mod}\,en} \qquad (2)$$

Three important characteristics of the general accessibility equation 2 are:

1. Values equal the ratio of the total number of opportunities to the total number of opportunity seekers. Therefore, if the number of jobs equals the number of workers, general accessibility equals one.

2. Values equal the potential for an opportunity seeker with specific skill sets to procure a job in a particular location. That is, they stratify the labor force population into like occupational groups and match the skill levels of residents to relevant employment opportunities.
3. Values are compared between varying modes of transportation, commute distances, and occupational categories.

Data sources and the role of GIS in the modeling process

The equations above require several data sets, including multi-modal impedance values or travel costs between residential and employment locations, employment of workers by both place of residence and place of work, and worker occupational skill sets. Most base demographic and spatial data required by the models were made available by national sources. A description of these data, their sources and applied function are provided below. ESRI's Arc/Info Geographic Information System (GIS) was used extensively for the spatial data analysis.

- TIGER/Line 1992 Maricopa County Street Network
  Road quality and spatial information provided with this transportation data set was used in a GIS to generate impedance costs. Speed limits, for example, were assigned to network segments based on road quality or function, which are identified by TIGER census feature class codes (CFCCs). This speed limit information was then used together with network segment lengths to calculate shortest network paths for traversing between residential and employment locations. A similar GIS network modeling technique was used by Wang to model commuting patterns in Chicago (2000).
- TIGER/Line 1992 Maricopa County Census Tracts and 1990 Census Summary Tape File 3A (STF3A)
  The areal units of analysis for this study were neighborhoods defined as individual census tracts. All demographic and employment data were allocated to and mapped by this geographic unit. The TIGER/Line files provide spatial boundary information and the summary tape file provided occupational and commuting characteristics of workers by place of residence. The means or transportation modes that workers used for work commutes were categorized into three levels of mobility. Workers who drove alone, carpooled, or used a motorcycle as their primary transportation to work were identified as having high mobility (MODE1). Those using public transportation were classified as moderately mobile (MODE2). And workers who either bicycled or walked to work were said to have low mobility (MODE3).
- 1990 Census Transportation Planning Package (CTPP)
  The Urban CTPP, released on CD-ROM in 1996 by the Bureau of Transportation Statistics, provides occupational and transportation characteristics of workers by place of employment. Part 2 of the CTPP Urban Element provides data by traffic analysis zone (TAZ)—a relatively small unit of geography—that

Table 1: Occupational Categories for Job Accessibility Calculations.

Category	Occupation
High skill (OCC1)	Executive, administrative, and managerial Professional specialty
Moderate to high skill (OCC2)	Administrative support including clerical Technicians and related support Protective services Sales
Low to moderate skill (OCC3)	Precision production, craft, and repair Transportation and material moving Service occupations, except protective and household Machine operators, assemblers, and inspectors Handlers, equipment cleaners, helpers, and laborers
Low skill (OCC4)	Farming, forestry, and fishing Private household services

were later aggregated into census tracts using a TAZ to Census Tract equivalency file.

- Bureau of Labor Statistics occupation by educational attainment
  In order to reduce redundancy and unnecessary complexity in the accessibility measures, it was necessary to group occupations into four main occupational categories based on skill level. Table 1 presents the four categories identified using the BLS occupation by educational attainment data set. Figure 2 charts the percentage of workers by occupational category and educational attainment.
- 1990 City of Phoenix Bus Book, Summer Edition
  The Metro Phoenix Bus Book, which includes origin and destination times of express and local bus routes, was used to calculate average velocities of public transit in metropolitan Phoenix. This measure was later used to calculate β in the exponential distance decay function discussed earlier in the paper.

Employment accessibility model results

The minimal public transit system is evidenced in the distribution of workers who commute by transit (Table 2). At three percent, the county's average is two points below the relatively stable national rate of people reaching work by transit (Bureau of Transportation Statistics 1999). Those who do utilize public transportation, however, reside predominantly in the central city (Figures 3 and 4). Further, captive public transit riders are more likely to be low to moderately skilled compared to the superior mobility status of high-skill workers (Figures 5 and 6).

Table 2: Mobility Categories and Average Modal Velocities for Job Accessibility Calculation.

Category	Means of Transportation to Work	Number of Workers[1] (%)	Average Speed[2]
High mobility (MODE1)	Drove alone Carpooled Motorcycle	898,286 (93.9)	35.3 mph
Moderate mobility (MODE2)	Bus or other public transportation	28,567 (3.0)	19.2 mph
Low mobility (MODE 3)	Bicycled or walked	40,333 (4.2)	7.8 mph

Data Sources: [1] US Bureau of the Census, STF3A, 1990; [2] Mode 1 velocity is based on speed limits and street network distances provided by Dynamap and 1992 TIGER/Line files; Mode 2 average velocity is based on bus routes identified in the Metro Phoenix Bus Book Summer Edition, 1990; Mode 3 average velocity is based on trip distance and length provided in National Transportation Statistics, 1996.

Figure 2.

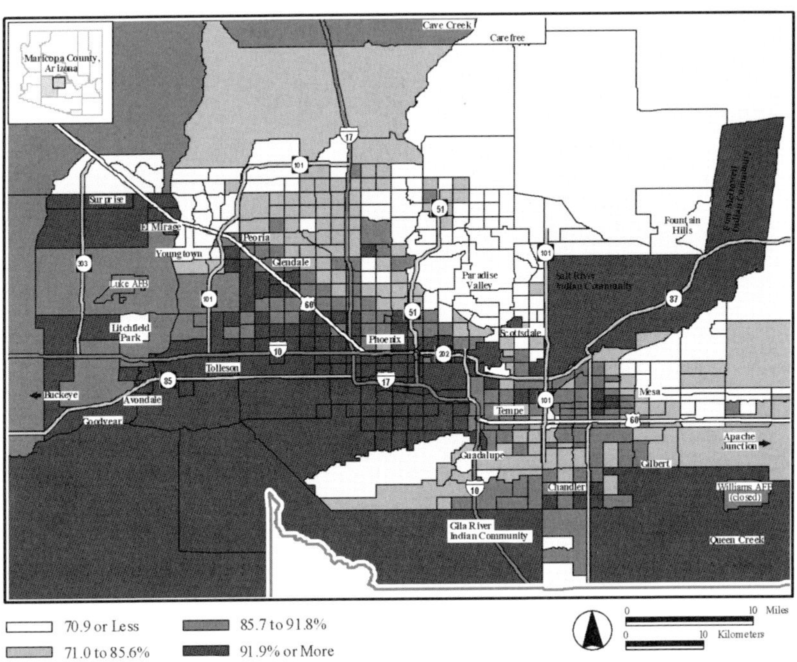

Figure 3.

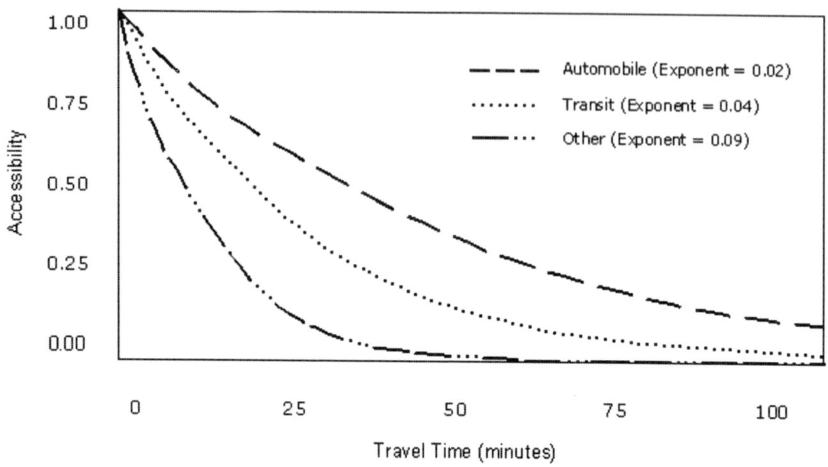

Figure 4.

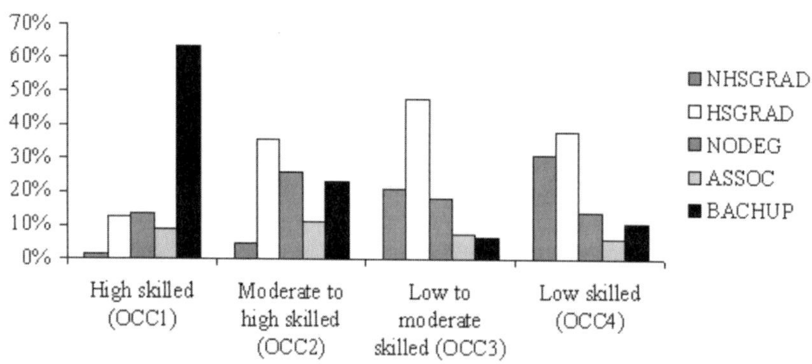

Figure 5.

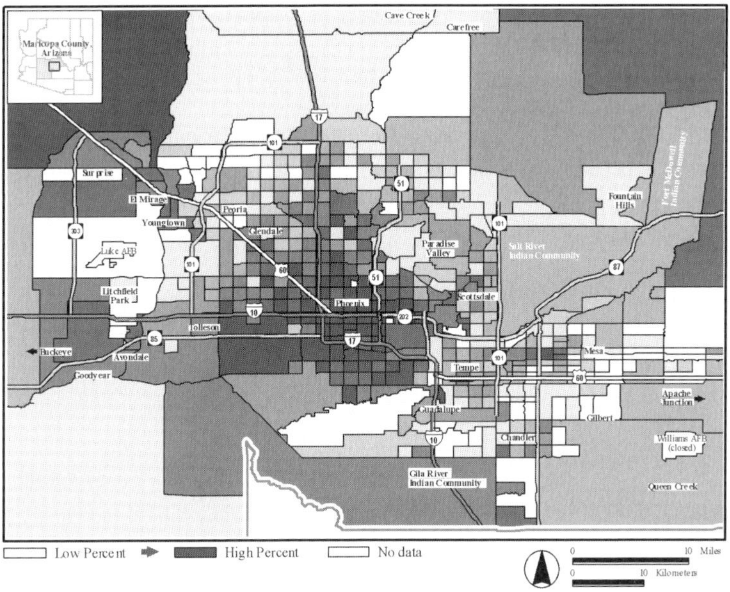

Figure 6.

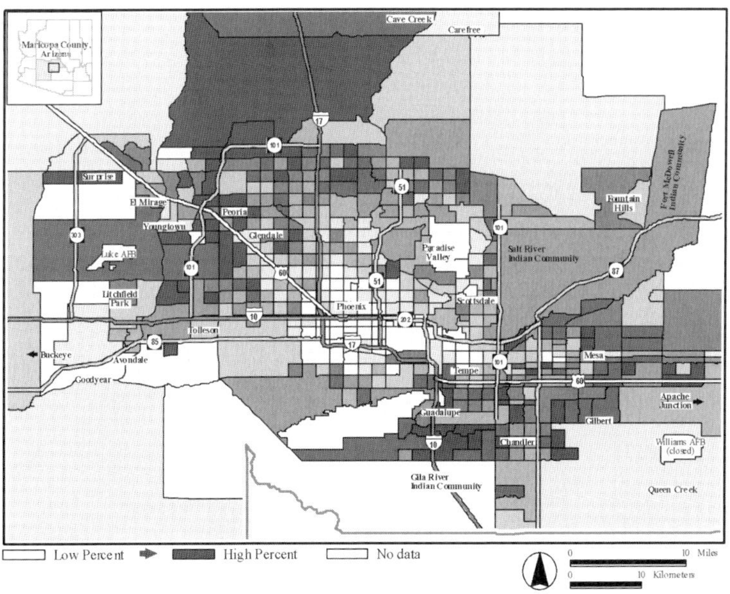

The combination of low skills and low mobility can exacerbate the economic hardships of central city workers. For example, the number of relevant job opportunities that can be accessed by workers with high mobility far surpasses the number of jobs that can be accessed by workers with lower mobility but similar skill sets. Figures 7 and 8 spatially present the distribution of job accessibility of low- to moderate-skill workers with different mobility potentials.

The relatively weak ratio of low-skill job availability to job demand is apparent in the central city (Figure 9). This pattern is in part due to the residential concentration of workers with low-skill sets, which amplifies the demand for low-skill jobs. This spatial distribution contrasts with the pattern of demand for high-skill jobs (Figure 10).

The ratio of high-skill job availability to job demand appears to be negatively correlated with neighborhoods occupied by high-skill workers. One anomaly is that of the city of Chandler where Intel and other high-technology manufacturers are located in close proximity to their workers, resulting in a concentration of high-skill workers with similar employment demands.

The distribution of high-skill jobs is more concentrated in the study region than low- to moderate-skill employment opportunities. The Phoenix central business district, where employment-to-population ratios are highest, comprises the greatest concentration of high-skill jobs (Figure 11). However, low- to moderate-skill jobs

Figure 7.

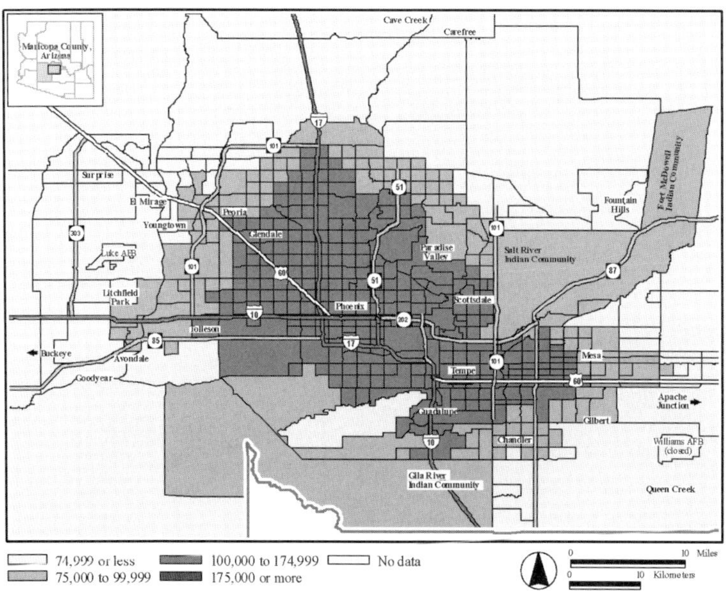

Figure 8.

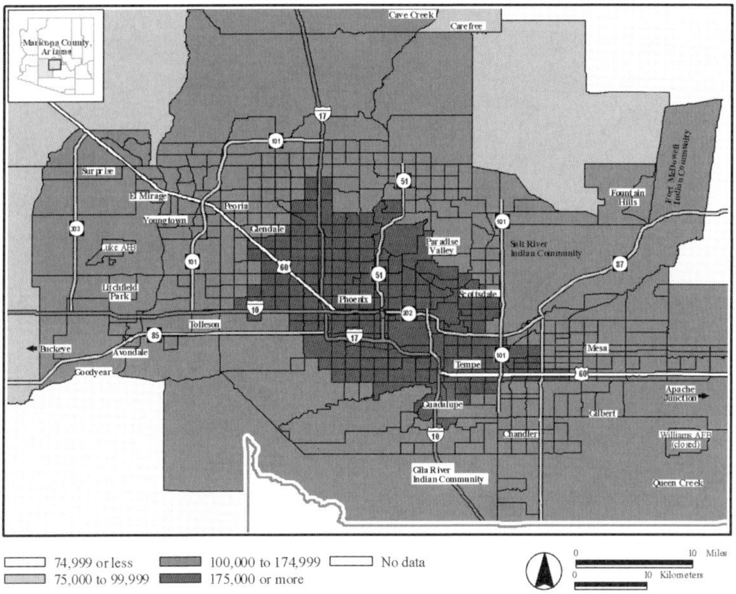

Figure 9.

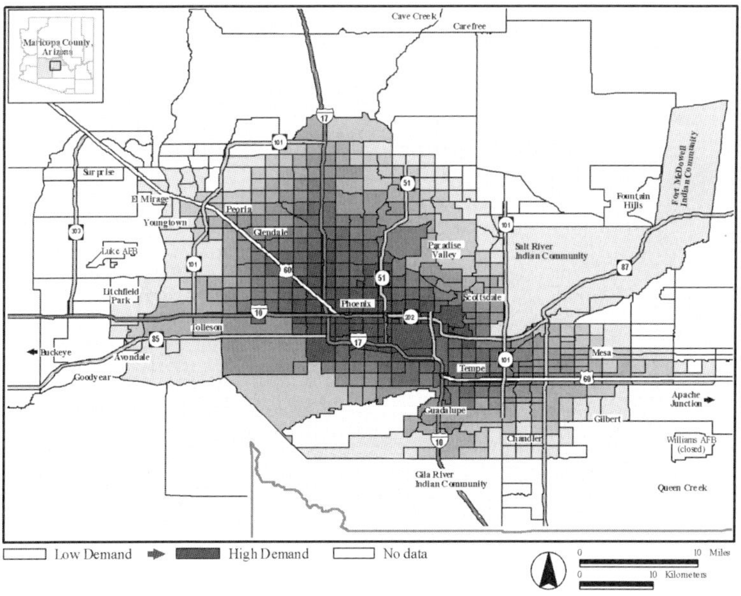

Figure 10.

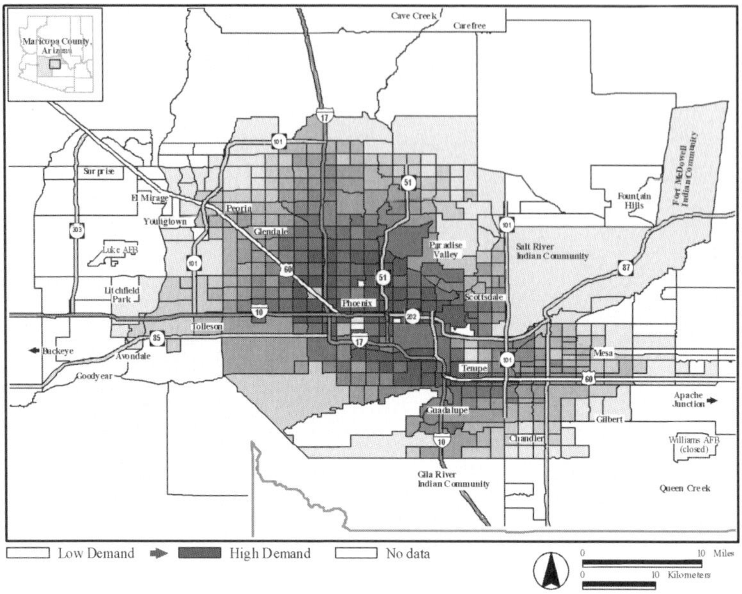

Figure 11.

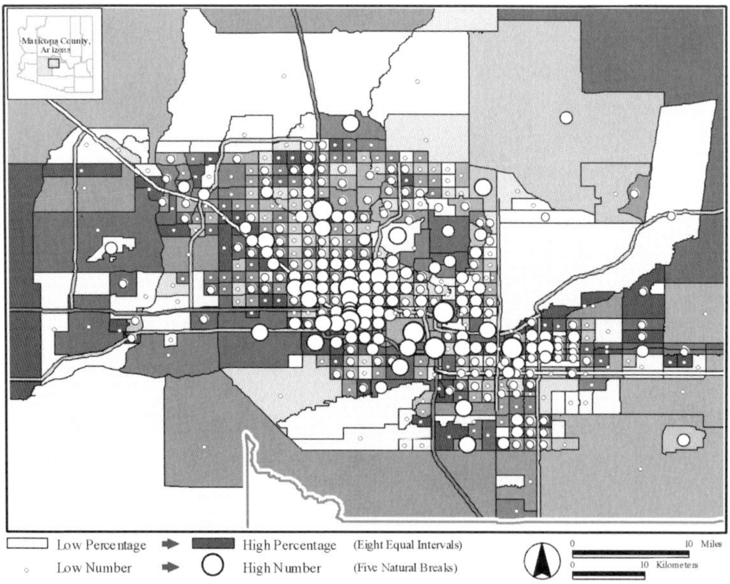

are evenly distributed throughout the landscape including areas with high residential growth (Figure 12). Note that the low- to moderate-skill employment demand in these areas are weak relative to the job opportunities.

Figures 13 and 14 show the spatial pattern of general accessibility (2) for low- to moderate-skill workers and high-skill workers respectively while accounting for all transportation modes. The visual pattern supports the spatial mismatch hypothesis in that employment accessibility for low- to moderate-skill workers and high-skill workers are low near where these populations reside.

In summary, the spatial distribution of job accessibility agrees with what was posited in the research—that a spatial mismatch exists between central city worker's residences and their relevant employment opportunities. Table 3 also supports the first hypothesis in that average job accessibility decreases with skill level. The next step is to determine whether these values are significantly and independently related to the economic performance of neighborhoods.

Figure 12.

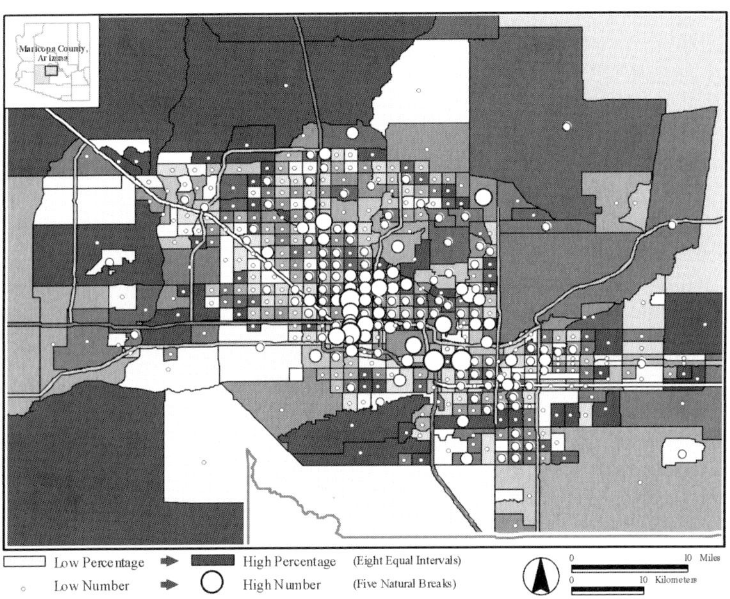

Table 3: Summary of Estimated General Accessibility Values by Occupation.

Occupation Category	General Accessibility*			
	Mean	Minimum	Maximum	SD
High skill (GAOCC1)	0.934	0.669	1.029	0.048
Moderate to high skill (GAOCC2)	0.923	0.614	1.008	0.047
Low to moderate skill (GAOCC3)	0.903	0.578	0.996	0.047
Low skill (GAOCC4)	0.845	0.494	0.955	0.050
All Skill Levels (GALL)	0.923	0.240	1.016	0.065

Note: Measures consider all transportation modes.

Figure 13.

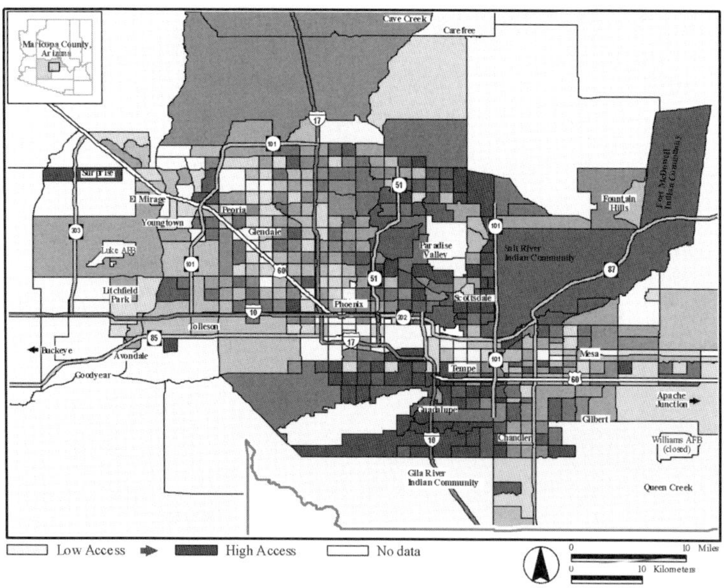

Figure 14.

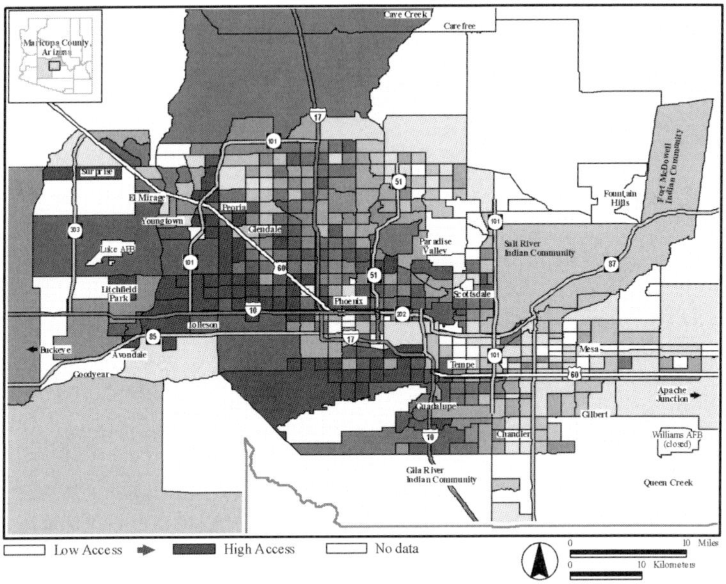

5. Fitting the Statistical Model

The present analysis followed a two-step procedure based in part on an approach recommended by Anderson and Gerbing (1988). In the first step, confirmatory factor analysis is used to develop two latent constructs—job accessibility and economic performance. Four indicators of each latent construct were identified from a pool of theoretically related variables. All latent and indicator variables were allowed to covary in a measurement model to determine whether the variables were indeed significantly and independently related. In step two, the measurement model was modified so that it came to represent the theoretical model in which job performance is "predicted" by the single latent variable job accessibility. The remainder of this section describes the process outlined above.

Table 4: Bivariate Correlations of Indicator Variables.

	NOINCOME	NOWORK	NOTINLF	CHPOVTY	NOVEHS	GAOCC3	BUS15LSS	WKR23VEH
NOINCOME	1							
NOWORK	0.892	1						
NOTINLF	0.846	0.885	1					
CHPOVTY	0.288	0.290	0.322	1				
NOVEHS	0.472	0.432	0.458	0.648	1			
GAOCC3	-0.301	-0.310	-0.296	-0.223	-0.289	1		
BUS15LSS	0.358	0.369	0.352	0.474	0.611	-0.209	1	
WKR23VEH	-0.282	-0.231	-0.213	-0.335	-0.503	0.151	-0.450	1

N = 441. All Correlations are significant at the 0.01 level (2-tailed).

Latent constructs: Job accessibility and economic performance

The variables job accessibility and economic performance are each difficult to capture using a single observed variable. This is evident by the numerous forms the variables have assumed in previous studies. Job accessibility, for example, has been measured as mean travel time, area ratio of jobs to workers, number of jobs within a 30-minute transit commute, and network and/or straight-line distance from residence to place of work (Ihlanfeldt 1992). Similarly, economic performance has been expressed as employment rate, employment probability estimates, labor force participation, and earnings.

Rather than allow a single measure to wholly represent one of these variables, this research utilized job accessibility and economic performance as latent constructs or factors to be inferred by observable or indicator variables. The factors, when integrated in a structural equation model, can support greater complexity by allowing second order factor relationships and information about indirect and direct variable effects. Such analyses cannot be performed in standard multivariate regression models.

Data in the study were analyzed using AMOS 4.0 structural modeling software and the models tested were covariance structure models. Standard deviations and bivariate correlations for the study's eight indicator variables are provided in Table 4.

Figures 15 and 16 present single-factor models for the latent constructs job accessibility and economic performance. Observed variables are represented as rectangles, latent constructs as ovals and measurement errors—the amount of error not explained by the latent construct—for the observed variables are represented as circles.

Figure 15.

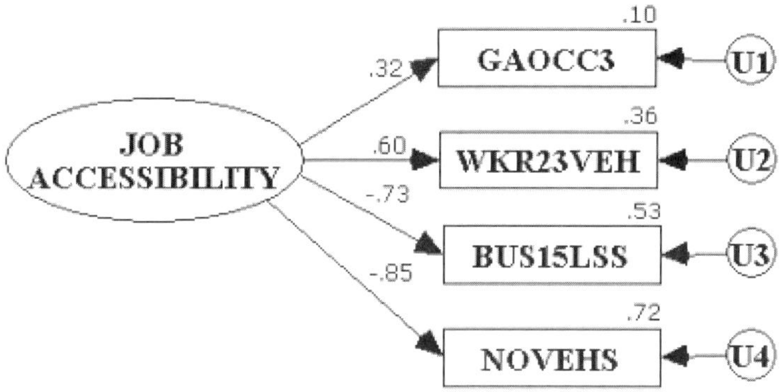

Figure 16.

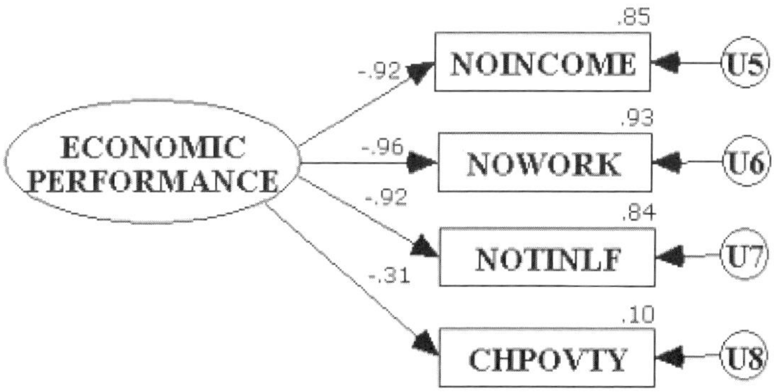

Table 5: Job Accessibility Single Factor Results.

Latent Construct	Observed Indicator	Standardized Estimates	Squared Multiple Correlations	Error Variance
Job Accessibility	GAOCC3	0.316	0.100	0.900
	WKR23VEH	0.599	0.358	0.642
	BUS15LSS	-0.725	0.526	0.474
	NOVEHS	-0.847	0.717	0.283
	Composite Reliability	0.729	Variance Extracted Estimate	0.425
	X^2 (df)	2.712 (2)	Probability level	0.258

N = 441

The observed, continuous variables were selected from a pool of theoretically related variables based on their ability to fulfill the assumption of multivariate normality.[5] The list of variables was further truncated based on their ability to uniquely load on their associated theoretical latent constructs—job accessibility or economic performance. Each variable's contribution to the model was initially tested using exploratory factor analysis then reaffirmed using confirmatory factor analysis.

The figures also present the standardized regression weights for each observed variable. All weights (i.e., factor loadings) have the predicted signs and all indicator variables either moderately or greatly load on their associated constructs. Job accessibility accounted for 71.7 percent of the variance of households with no vehicles (NOVEHS), followed by workers that commute by bus (BUS15LSS, 52.6 percent), two-worker households with three vehicles (WKR23VEH, 35.8 percent), and general accessibility of low- to moderately-skilled workers (GAOCC3, 10 percent) (Table 5). Economic performance accounted for 93.0 percent of the variance of the rate of population that did not work in 1989 (NOWORK), followed by neighborhood rates of no 1989 income (NOINCOME, 85.3 percent), labor force participation (NOTINLF, 84.3 percent) and child poverty (CHPOVTY, 9.9 percent) (Table 6).

[5] The Kolmogorov-Smirnov goodness-of-fit test was used to test whether the empirical distribution of the observations was consistent with a random sample drawn from a normal distribution.

Table 6: Economic Performance Single Factor Results

Latent Construct	Observed Indicator	Standardized Estimates	Squared Multiple Correlations	Error Variance
Economic Performance	NOINCOME	-0.924	0.853	0.147
	NOWORK	-0.964	0.930	0.070
	NOTINLF	-0.918	0.843	0.157
	CHPOVTY	-0.315	0.099	0.901
	Composite Reliability	0.884	Variance Extracted Estimate	0.578
	X^2 (df)	4.736 (2)	Probability level	0.094

N = 441

Both constructs exceed the minimal acceptable composite reliability of 0.600 and have nonsignificant chi-squares marked by corresponding p values above 0.5.[6] The relatively low variance extracted estimate for job accessibility (0.425) indicates that variance due to measurement error exceeds variance captured by the factor. However, this parameter is conservative and, given the other satisfactory indices, may not be wholly indicative of factor reliability.

Initial measurement model

An initial measurement model (M_i) was constructed using the latent factors and indicators outlined above (Figure 17). The covariance between the latent factors—represented by a curved arrow—was estimated using the maximum likelihood method (ML) and the chi-square value for the model was statistically significant x^2 (19, $\underline{N}$ = 441) = 249.13 (Table 7). In addition to the unsatisfactory chi-square, a number of other results indicated that the model was not appropriately fit. The squared multiple correlation of CHPOVTY suggests that economic performance accounts for only 10 percent of the variance in child poverty rates. The model, therefore, does not explain the remaining 90 percent of the variation of this indicator.

Revised measurement models

Goodness of fit indices for the initial and respecified models are presented in Table 7. Their associated measurement model diagrams are shown in Figure 18.

[6] When the proper assumptions are met (large sample, multivariate normal distribution), the chi-square test provides a statistical test of the null hypothesis that the model fits the data. Therefore, a good model is one with a relatively small or nonsignificant chi-square value.

Figure 17.

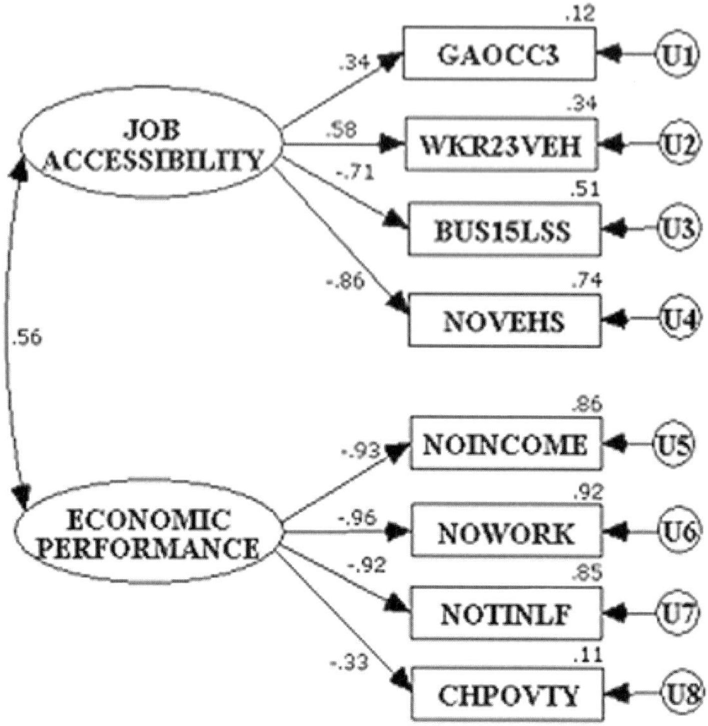

Table 7: Measurement Models Goodness of Fit Indices.

	Measurement Model		
Index	M_i	M_1	M_2
Chi-Square	249.133	30.610	11.242
df	19	8	6
X^2 Probability	0.000	0.000	0.081
NFI	0.887	0.981	0.993
CFI	0.895	0.986	0.997
PR	0.679	0.533	0.400

N = 441; NFI = normed fit index; CFI = comparative fit index; PR = parsimony ratio.

Figure 18.

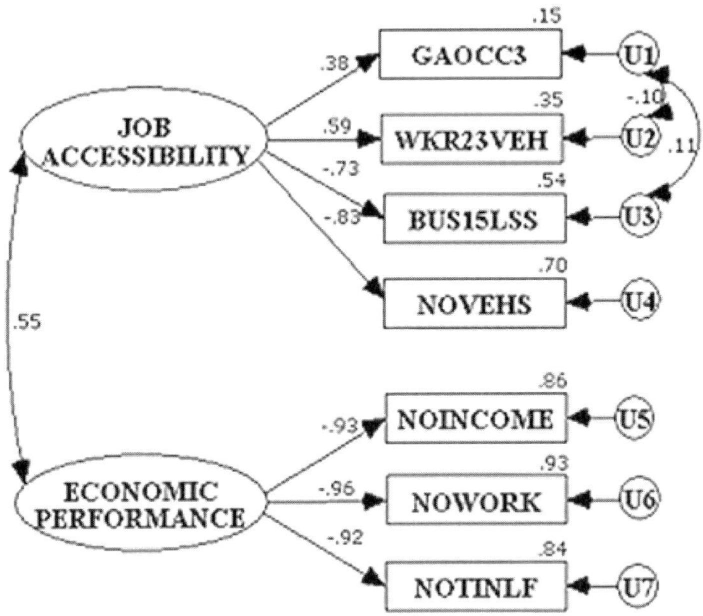

Dropping both CHPOVTY and NOVEHS from the analysis considerably improved the first revised model (M_1) marked by a lower chi-square value and greater model fit.

Two goodness of fit indices were used in this analysis: the normed fit index, or NFI, and the comparative fit index, or CFI. Both values may range from 0 to 1, where 0 represents the goodness of fit associated with a model with all uncorrelated variables, and 1 represents the goodness of fit associated with a model that perfectly reproduces the original covariance matrix (Bentler 1980; Bentler 1989).

The parsimony or relative simplicity of the model was also tested using the parsimony ratio or PR (James 1982). The ratio equals the degrees of freedom for the model divided by the degrees of freedom of the null model (the null model predicts no relationships between any of the study's variables) and ranges from 0 to 1. Therefore, an upper value of 1 indicates the most parsimonious model possible because it makes the fewest assumptions between the variables.

Adding error covariances between GAOCC3 and WKR23VEH, and GAOCC3 and BUS15LSS, in the second revised model (M_2) allowed the chi-square to be rejected at the 95 percent level. Although the parsimony ratio is low, indicating the complexity of the model is high compared to the explanatory power, the other satisfactory indices were encouraging enough to accept this second revised model as the final measurement model.

Figure 19.

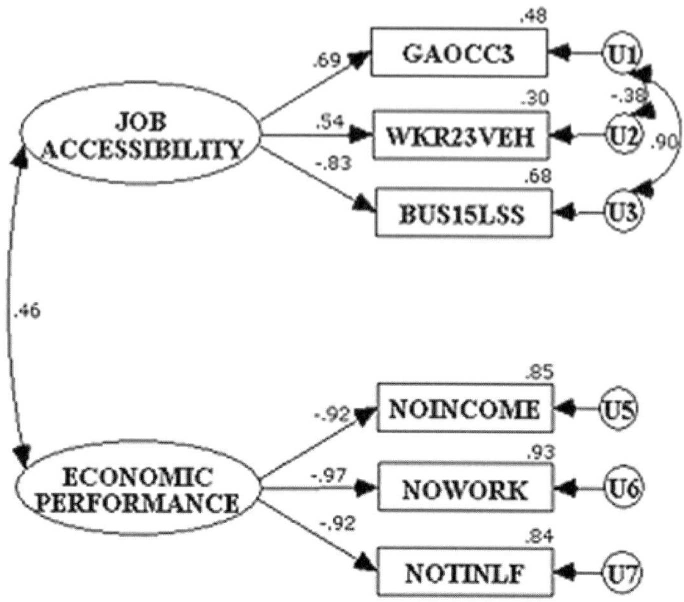

The final measurement model was then modified to show the causal path associated with the hypothesis that job accessibility is related to economic performance. This is represented in Figure 20 as a single headed arrow drawn between the latent constructs. Because it is a simple structural model with a single independent and dependent latent variable, the model fit indices are identical to the final measurement model.

The direct, indirect, and total effects of the variables are presented in Table 8. The final model indicates that over 46 percent of the variance of economic performance is accounted for by job accessibility. Therefore, greater access to jobs has a significant effect on a community's economic situation. BUS15LSS, GAOCC3, and WKR23VEH account for over 48 percent of the factor variance of job accessibility. NOTINLF, NOWORK, and NOINCOME account for almost 88 percent of the factor variance of economic performance

6. Conclusions and Recommendations

This research has explored (a) forces that influence spatial mismatch, (b) refined measures of job accessibility, and (c) a statistical model that relates these measures to the economic performance of central city neighborhoods in metropolitan

Figure 20.

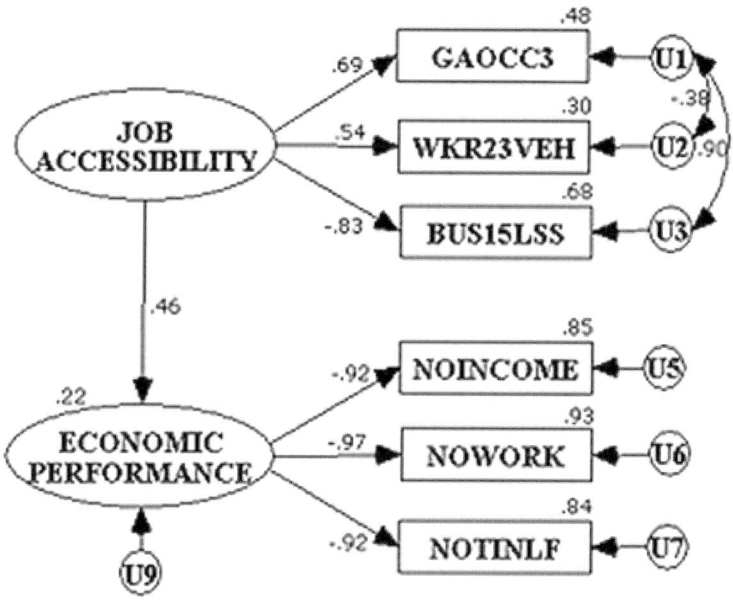

Table 8: Structural Equation Model Results.

Latent Construct	Observed Indicator	Standardized Estimates	Squared Multiple Correlations	Error Variance
Economic Performance	NOINCOME	-0.924	0.854	0.146
	NOWORK	-0.965	0.931	0.069
	NOTINLF	-0.917	0.841	0.159
	Composite Reliability	0.955	Variance Extracted Estimate	0.875
Job Accessibility	GAOCC3	0.694	0.482	0.518
	BUS15LSS	-0.826	0.682	0.318
	WKR23VEH	0.545	0.297	0.703

N = 441

Phoenix. Thus it is fitting to end this paper with a brief discussion of why planners should operationalize job accessibility measures in planning practice and, more specifically, how these measures can be used to improve social equity in transportation systems and enhance the link between land use, environmental, and transportation planning.

Recent policy changes, existing environmental justice mandates, and smart growth and sustainable development principles exemplify reasons why planners should operationalize measures of accessibility. The U.S. Department of Transportation's Transportation Equity Act (TEA-21) program, for example, has a significant job accessibility component. Activated in 1999, the component is commonly referred to as the Job Access and Reverse Commute or Welfare to Work program and assists states and municipalities in developing new or expanded transportation services for connecting low-income persons to jobs and other employment related services. Toward this end, job accessibility values can be applied to identify communities in need of transportation services. Once identified, a collaborative effort could be made by metropolitan planning organizations, transportation providers, affected communities, and their stakeholders to develop a project that serves the community. In 2000, Congress appropriated $75 million for such programs.

Planners could also operationalize accessibility to promote environmental justice in transportation. In addition to TEA-21, there are laws, regulations, and policies that require the consideration of minority and low-income populations. For instance, Title VI of the Civil Rights Act of 1964 mandates that "no person in the United States shall be denied the benefits of any program or activity receiving Federal financial assistance." Further, a 1994 Presidential Executive Order directed every federal agency to consider the effects of all programs, policies, and activities on disadvantaged communities. In order to comply with Title VI and the executive order, metropolitan planning organizations need to enhance their analytical capabilities to ensure that long-range transportation plans and investments are fairly distributed. Hence, job accessibility measures could be incorporated in urban modeling and alternative plan analysis to determine whether prospective initiatives work toward accomplishing these goals. UrbanSIM, an urban modeling application developed by Paul Waddell and the University of Washington, is an example of an effective model that addresses the interactions between housing and labor markets using job accessibility values (Waddell 1998; 2001).

Finally, planners would benefit by using job accessibility measures to help forward some of the most basic principles outlined in smart growth, New Urbanism, sustainable development, and related anti-sprawl initiatives. America's transportation infrastructure since the 1950s has prioritized highway building, fostering an auto-dependent society. The initiatives mentioned above attempt to remedy the deep structural imbalances caused by such a society and the sprawling urban form that supports it. Job accessibility measures could be used as performance indicators of progress toward smart growth.

The Netherlands, for example, has made considerable headway in applying job accessibility measures. They have classified various locations within cities by accessibility levels that help planners and local area citizens support prospective businesses that match the accessibility profiles of their neighborhoods (Cervero, Rood, and Appleyard 1995).

In conclusion, job accessibility values, when measured appropriately, can be implemented in ways that advance citizen participation, equitable transportation, and overall efficiency in urban design.

References

Abbitt, Jerry W. 1990. *History of transit in the Valley of the Sun: A history of public transportation in Phoenix, Arizona, 1887–1989*. Phoenix: The City of Phoenix Transit System.

Anderson, J. C., and D. W. Gerbing. 1988. Structural equation modeling in practice: A review and recommended two-step approach. *Psychological Bulletin* 103:411–423.

Arnott, R. 1998. Economic theory and the spatial mismatch hypothesis. *Urban Studies* 35 (7):1171–1185.

Bauder, H., and E. Perle. 1999. Spatial and skills mismatch for labor-market segments. *Environment and Planning A* 31:959–977.

Beatley, Timothy. 1994. *Ethical land use: Principles of policy and planning*. Baltimore: The Johns Hopkins University Press.

Bendick, Marc Jr., Charles W. Jackson, and Victor A. Reinoso. 1994. Measuring employment discrimination through controlled experiments. *Review of Black Political Economy* 23:25–48.

Bentler, P. M. 1989. *EQS structural equations program manual*. BMDP Statistical Software, Los Angeles.

Bentler, P. M. and D. G. Bonett. 1980. Significance tests and goodness-of-fit in the analysis of covariance structures. *Psychological Bulletin* 88:588–606.

Boswell, T. D., and A. D. Cruz Baez. 1997. Residential segregation by socioeconomic class in metropolitan Miami: 1990. *Urban Geography* 18 (6):474–496.

Bureau of Transportation Statistics. 1999. *Transportation statistics annual report 1999*. Washington, D.C.: U.S. Department of Transportation Bureau of Transportation Statistics.

Burns, E. K., and P. Gober. 1998. Job linkages in inner city Phoenix. *Urban Geography* 19 (1):12–23.

Center for Business Research. 2000. *Arizona economic profile*. Tempe, Ariz.: Arizona State University, L. William Seidman Research Institute.

Cervero, Robert, Timothy Rood, and Bruce Appleyard. 1995. *Job accessibility as a performance indicator: An analysis of trends and their social policy implications in the San Francisco Bay area*. Berkeley, Calif.: University of California Transportation Center.

Cuomo, Andrew. 1999. *New reports document discrimination against minorities by mortgage lending institutions*. Washington D.C.: Office of Public Affairs.

Current Population Survey. 2000. *March current population survey*. Washington D.C.: U.S. Census Bureau/U.S. Department of Commerce.

Darden, J. T., J. G. Bagaka, and Ji Shun Jie. 1997. Racial residential segregation and the concentration of low- and high-income households in the 45 largest U.S. metropolitan areas. *Journal of Developing Societies* 13 (2):171–194.

Ellwood, David T., ed. 1982. Teenage unemployment: Permanent scars of temporary blemishes. In *The youth labor market problem: Its nature, causes, and consequences*, edited by R. B. Freeman and D. A. Wise. Chicago: University of Chicago Press.

Gammage, Grady Jr. 1999. *Phoenix in perspective*. Tempe, Ariz.: Arizona State University, Herberger Center for Design Excellence.

Hughes, M. A. 1990. Formation of the impacted ghetto: Evidence from large metropolitan areas, 1970–1980. *Urban Geography* 11 (3):265–284.

———. 1991. Employment decentralization and accessibility: A strategy for stimulating regional mobility. *Journal of the American Planning Association* 57 (3):288–298.

Ihlanfeldt, K. 1992. *Job accessibility and the employment and school enrollment of teenagers*. Kalamazoo, Mich.: W.E. Upjohn Institute for Employment Research.

Ihlanfeldt, K. R., and M. V. Young. 1999. The spatial distribution of black employment between the central city and suburbs. *Economic Inquiry* 34 (4):693–707.

James, L. R., S. A. Mulaik, J. M. Brett. 1982. *Causal analysis*. Beverly Hills: Sage.

Kain, John. 1968. Housing segregation, negro employment and metropolitan decentralization. *The Quarterly Journal of Economics* 82 (2):175–197.

Kasarda, John D. 1990. Structural factors affecting the location and timing of urban underclass growth. *Urban Geography* 11 (3):234–264.

———. 1994. Industrial restructuring and changing job locations. Paper presented at the annual meeting of the Population Association of America at Miami, Florida.

Knox, Paul L. 1990. The new poor and a new urban geography. *Urban Geography* 11 (3):213–216.

Kunstler, James Howard. 1993a. Are we zoned for destruction? *The New York Times*, Monday, August 9, 1993, 196–198.

———. 1993b. *The geography of nowhere*. New York: Simon & Schuster.

Lewis, Oscar. 1966. The culture of poverty. *Scientific American* 215 (4): 19–22.

Luckingham, Bradford. 1994. *Minorities in Phoenix: A profile of Mexican American, Chinese American and African American communities 1860–1992*. Tucson: The University of Arizona Press.

Marcuse, Peter. 1997. The enclave, the citadel, and the ghetto: What has changed in the post-Fordist U.S. city. *Urban Affairs Review* 33 (2):228–264.

Massey, Douglas S., and Nancy A. Denton. 1993. *American apartheid: Segregation and the making of the underclass*. Cambridge, Mass.: Harvard University Press.

McKenna, C. J. 1996. Education and the distribution of unemployment. *European Journal of Political Economy* 12 (1):113–32.

McLafferty, S. 1996. Spatial mismatch and employment in a decade of restructuring. *Professional Geographer* 48 (4):420–431.

Meiklejohn, Susan Turner. 2000. *Wages, race, skills and space*. New York: Garland Publishing.

Morrison Institute for Public Policy. 2000. *Hits and misses: Fast growth in metropolitan Phoenix*. Tempe, Ariz.: Arizona State University, Morrison Institute for Public Policy.

Murray, Charles. 1984. *Losing ground: American social policy 1950–1980*. New York: Basic Books.

Neckerman, Kathryn M., and Joleen Kirshenman. 1991. Hiring strategies, racial bias, and inner-city workers. *Social Problems* 38 (4):801–815.

Norquist, John O. 1998. *The wealth of cities: Revitalizing the centers of American life.* Reading, Mass.: Addison-Wesley.

Roseland, Mark. 1998. *Toward sustainable communities: Resources for citizens and their governments.* Gabriola Island, BC: New Society Publishers.

Schill, Michael H., and Susan M. Wachter. 1995. The spatial bias of federal housing law and policy: Concentrated poverty in urban America. *University of Pennsylvania Law Review* 143:1285–1379.

Shen, Q. 1998. Location characteristics of inner city neighborhoods and employment accessibility of low-wage workers. *Environment and Planning B* 25 (3):345–365.

Sheppard, Eric. 1990. Ecological analysis of the "urban underclass": Commentary on Hughes, Kasarda, O'Regan and Wiseman. *Urban Geography* 11 (3):285–297.

Stoll, M. 1999. *Race, space and youth labor markets.* New York: Garland Publishing.

Summers, A., P. Cheshire, and S. Lanfranco. 1993. *Urban change in the United States and Western Europe: Comparative analysis and policy.* Lanham, Md.: Urban Institute Press.

Turner, Susan C. 1997. Barriers to a better break: Employer discrimination and spatial mismatch in metropolitan Detroit. *Journal of Urban Affairs* 19 (2):123–141.

U.S. Census Bureau. 1990. *Census of population and housing characteristics.* Washington D.C.: U.S. Bureau of the Census.

U.S. Department of Transportation. 1997. *BTS annual report 1997.* Washington, D.C.: Bureau of Transportation Statistics.

Wacquant, L. 1993. Urban outcasts: Stigma and division in the black American ghetto and the French urban periphery. *International Journal of Urban and Regional Planning* 17 (3):366–83.

Waddell, Paul. 1998. An urban simulation model for integrated policy analysis and planning: Residential location and housing market components of UrbanSIM. Paper read at 8th World Conference on Transport Research, July 12–17, 1998, at Antwerp, Belgium.

———. 2001. Towards a behavioral integration of land use and transportation modeling. Paper read at 9th International Association for Travel Behavior Research Conference, January 2001 at Queensland, Australia.

Wang, F. 2000. Modeling commuting patterns in Chicago in a GIS environment: A job accessibility perspective. *Professional Geographer* 52 (1):120–133.

Wilson, William Julius. 1987. *The truly disadvantaged: The inner city, the underclass, and public policy.* Chicago: University of Chicago Press.

———. 1996. *When work disappears: The world of the new urban poor.* New York: Vintage Books.

World Bank. 2001. *World development indicators database.* Washington, D.C.: World Bank.

Maricopa Association of Governments (MAG) Review of the Maricopa Region and the MAG Socioeconomic Projection Process

Rita Walton, Mike Corlett, Cathy Arthur, and Anubhav Bagley

Introduction

Planners, developers, real estate brokers, and others involved in real estate development can give you an educated guess about what land in a region is likely to be built, and what land is or is not likely to be built on, based on their knowledge of land costs, zoning factors, infrastructure availability, water and utility access, site access and visibility, environmental conditions, city development review practices, and local market conditions.

Quantifying opinions about future growth patterns is certainly impractical on a massive scale such as that of Maricopa County, where there are over a million parcels covering more than 9,000 square miles and land development forecasts are needed to place over 1.5 million additional residents over the next 25 years. Even if land-use forecasts are needed at five-year intervals over the next four decades, urban growth models should still take into consideration the factors that interest developers.

Subarea Allocation Model—Information Manager (SAM-IM) is a rule-based urban growth model. It simulates both short-term and long-term urbanization of a region by reacting to any set of factors and conditions that the planner wishes to express. The model is embedded in a Geographic Information System (GIS)—it runs on ArcView using ArcView's Spatial Analyst extension. The concept of the model is entirely GIS-oriented—all of the data that drives the model, whether it be existing land-use distributions, future market conditions, adopted planned land use, developments already approved and underway, or local land conditions, are expressed geographically in the form of ArcView shape files. Anything that can be expressed geographically can be taken into consideration in the model.

This paper describes the SAM-IM suite of modeling applications that have been developed to support land-use forecasting at the Maricopa Association of Governments (MAG), the Metropolitan Planning Organization (MPO) and regional planning agency for the Phoenix metropolitan area.

GIS at MAG

MAG is responsible for regional land use, transportation, and air quality modeling for the Phoenix metropolitan area. Like other regional governments with similar MPO responsibilities, MAG seeks to improve the performance of these models, thereby making the regional planning process more effective. GIS has played a major role at MAG for the past 12 years. This technology has significantly enhanced MAG's ability to manage dozens of planning databases, including those describing census statistics, employment inventories, land use, general plans, parcels, building permits, and highway infrastructure, which all share one common characteristic—they are essentially spatial in nature. GIS helps MAG assume an important responsibility in the region—the role of regional information agency for public planning.

GIS has capabilities that run far beyond just being a platform for planning and mapping. In recent years, MAG has been developing new classes of planning models that work directly from GIS databases. More importantly, these models tap a wide variety of powerful spatial analysis methods that are found only in GIS and can not be matched by older custom-written programs developed over the last three decades. In the long term, this new class of GIS-based models offers the potential for replacing older, nonspatial planning models.

This paper describes one such model: the Subarea Allocation Model (SAM), developed and run at MAG. SAM forecasts land use and development throughout the MAG planning region. It does this by simulating factors that influence the value of land and the likelihood that land will be built on, based on those factors. It observes planning policies—general plan designations controlling the use of land (approved by municipalities in the region), for example.

The land-use, population, and socioeconomic modeling at MAG is based on a three-tier modeling process. The first tier is a demographic model that is used to produce county control totals. The second tier involves using a spatial interaction model to allocate the county control total population and employment to subregions. The third tier allows for the allocation of the subregional population to smaller areas drawing upon GIS representation of land-use plans and local policies of MAG member agencies.

The first-tier model is a county level model. In accordance with Executive Order 95-2, the preparation of county- and state-level population projections is the responsibility of the Arizona Department of Economic Security (DES). This model is a demographic model, projecting births, deaths, and net migration in each county for a 50-year horizon. This model incorporates population by age and sex, birth rates, death rates, and net migration trends. The model takes into account short-term economic conditions but not long-range employment trends.

Figure 1. The SAM-IM simulation generated this forecast of growth and development in Maricopa County over the next 20 years. Shown in gray are existing built uses (in 1995). Shown in black are new developments projected to occur by the year 2020. The proposed 2020 freeway system is also shown.

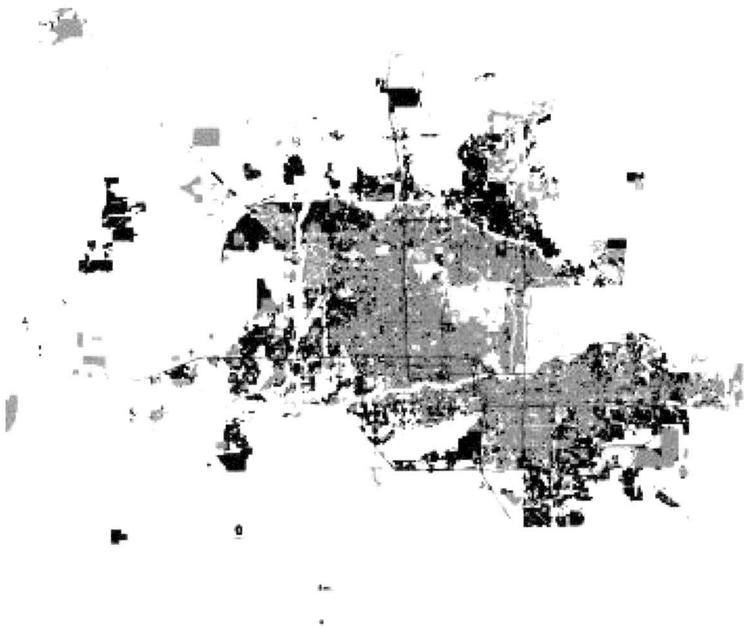

For the second tier process, MAG is using DRAM/EMPAL. DRAM (Disaggregated Residential Allocation Model) and EMPAL (EMPloyment Allocation Model) forecast household location and employment location, projecting the spatial patterns of households and employment in the MAG region. The forecasting procedure starts with regional trends, transportation facility descriptions, and data on the current location of employment by sector. This information is then used to project the future location of households. The projections are done for five-year intervals. Each five-year step begins with the EMPAL model to project employment by sector by zone. DRAM modeling to project households by income category follows the EMPAL run for that time period.

The third-tier SAM allocates population and employment from Regional Analysis Zones (RAZ) to one-acre grids which are then aggregated to Traffic Analysis Zones (TAZ). SAM generates simulations of how Phoenix will grow as a region over time. Its current role is to bridge the gap between the spatial allocation model, DRAM/EMPAL, and EMME/2, MAG's transportation model that works on much more detailed geographies. So, SAM's official role is as an allocation model in that it disaggregates land-use forecasts generated by DRAM/EMPAL for large statistical areas to smaller TAZs that are needed to drive the transportation model.

Model Integration

The sequence of land-use and transportation models run at MAG is recursive. DRAM/EMPAL requires assumptions about future travel times between RAZs across the transportation network in order to generate forecasts of land use. These travel time forecasts are available from the transportation model. In turn, the EMME/2-based transportation model requires forecasts of future land use in order to generate estimates of travel times between RAZs. This has led to the development of a MAG feedback loop, whereby the DRAM/EMPAL and EMME/2 model chains are executed a number of times, feeding the results of one model back into the other, until closure is achieved. This is necessary to ensure that the results generated by one model are consistent with the assumptions made by the other.

From a modeling standpoint, the integration these of two independently built models presents two significant database issues:

- Geographic detail: DRAM/EMPAL and EMME/2 operate at two different levels of geography. With roughly a 10-to-1 ratio between RAZ and TAZ geographies, the first severe problem that arises is one of data disaggregation.
- Secondary, or derived, variables: DRAM/EMPAL does not supply all of the land-use variables that are needed by EMME/2. DRAM/EMPAL generates estimates of households, for example. The EMME/2 transportation model requires estimates of resident population and dwelling units as well as households. So there is need for a mechanism that can supply secondary variables needed by the transportation model. Some of these variables can be easily derived from the DRAM/EMPAL forecasts themselves (e.g., dwelling unit estimates can be generated from household estimates based on appropriate assumptions about vacancy rates). Other variables, such as total enplanements at Phoenix Sky Harbor International Airport, must come from other secondary forecast sources

Subarea Allocation Models

The Subarea Allocation Model (SAM) was first posed as a land-use disaggregation model. The objective was to develop an automated procedure to bridge the geographic gap between DRAM/EMPAL forecasts generated for 147 RAZs and the needs of the transportation model which is driven by a 1,549 TAZ system. Also, the automated procedure had to inject other variables needed by the transportation model, some of which are highly correlated with DRAM/EMPAL forecasts and therefore should be derived from them.

Figure 2. SAM-IM provides a bridge between forecasts generated by DRAM/EMPAL at a relatively coarse geographic scale to the transportation model EMME/2, which operates with much greater geographic detail.

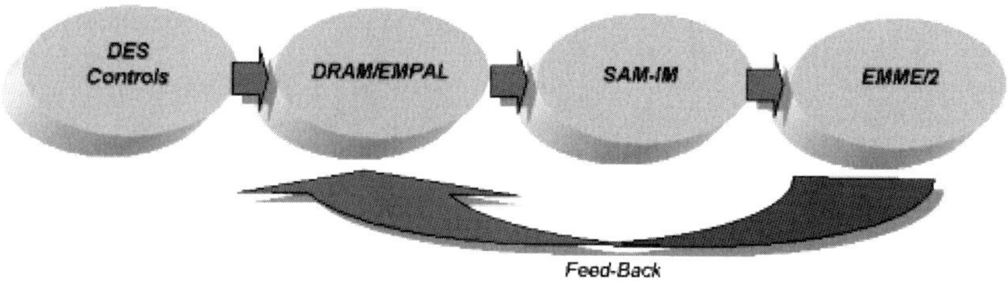

Before embarking upon the development of SAM, MAG's level of investment in GIS capability was sufficiently advanced to support the model's requirements. MAG had developed existing land use covers for all of Maricopa County, and it had assembled and unified the general plan covers adopted by municipalities throughout Maricopa County. MAG had also mounted a number of other efforts germane to SAM-type model concepts. For example, MAG was tracking, on GIS, active developments throughout the region as well as planned developments in various stages of the approval process. Retirement communities had been mapped, redevelopment district coverages were available and MAG had a capability to represent future EMME/2 transportation networks in GIS. So, in short, the database support needed for a SAM-type model was already available

The original SAM model was implemented as a set of ARC/INFO AMLs running on UNIX, using the ARC/INFO GRID module to convert feature coverages to grids. Much of the logic associated with allocating growth capitalized on spatial analysis functions available in ARC/INFO GRID. In fact, MAG had participated in the development of a grid-based model with a 40-acre resolution in the late 1980s, but ARC/INFO GRID allowed MAG to consider modeling at a much finer resolution.

The implementation of SAM was completed in 1996–97 and was used for the purposes of generating land-use forecasts for the region for five-year intervals, 2000 through 2020. These forecasts were reviewed with the member governments and were ultimately adopted by the Regional Council in 1997. Some of the key characteristics associated with the SAM model are:

- Growth model: SAM is a growth model; the entire forecast is not allocated, only the growth implied with it is allocated.
- Iterative model: It is an iterative model; the forecast generated for a target year (say 2005) is taken as a "given preexisting condition" for the purposes of generating a forecast for a subsequent target year (say 2010). Growth is "accumulative." Consequently, a forecast generated for a future target year is assured of being consistent with the forecast already made for an earlier target year.
- Cell resolution: SAM is operated countywide with a grid cell size of 220 feet, or approximately 1.1 acres.

- Allocation variables: The SAM model allocated households along with the five classes of employment forthcoming from the DRAM/EMPAL model. In addition, the SAM model allocated a number of other variables (group quarters, motels, and mobile home sites) to individual sites within the county directly from countywide control total forecasts.
- Site location considerations: Different site location characteristics were employed depending on the allocation variable. The allocation mechanism for housing and employment classes was represented by measures of proximity to existing infrastructure (represented by the future transportation network), proximity to existing development, and distance from the urban area. The "site location" characteristics for other types of allocation variables were different. Proximity to freeways, for example, was found to be a prime determinant of motel/hotel development.

By 1997, ESRI had released the Spatial Analyst extension to ArcView, which provide ArcView with virtually the same desktop capabilities for raster geographies as does ARC/INFO GRID. When considering the future development and enhancements to SAM, it became apparent that the entire model should be re-implemented in ArcView. Many reasons contributed to this decision, including:

- ArcView facilitated the migration from a UNIX environment to a Windows environment; other applications common to Windows environments, such as spreadsheet and database applications, would be available for analysis and forecasting work.
- ArcView already offered an extensive and robust set of menu items, commands, and functions for managing and mapping land-use data sets that would have otherwise had to be built into SAM in ARC/INFO.
- ArcView operating in a Windows environment integrates well with Visual BASIC; so the modeling system could benefit from customized programs and forms.
- Avenue, ArcView's scripting language, appeared to be better and faster in operation than AML.
- ArcView offered an excellent graphics user interface (GUI) which facilitated the operation of the model.
- ArcView offered improved mapping capabilities for grids.

The new model has been dubbed the (SAM-IM). The reimplementation in ArcView provided the opportunity to significantly enhance the model application itself, including these features:

- SAM-IM is "configurable." Users can modify the land-use database structures, and add or delete land-use codes to the coding scheme, without any change to the software. Users can also change the definitions of the forecast allocation variables themselves.
- SAM-IM provides the users with land-use database editors. These editors give users an easy way to modify land-use data sets, whether they represent existing land use, planned land use, active development projects, or whatever land-use source data is of interest.

- A new site-scoring module gives users a battery of utilities especially designed to create grids that reflect various aspects of site suitability of land. These menu-driven utilities let users evaluate land—measure its proximity to roadways and other types of infrastructure, measure distances from other features, automatically acquire location attributes from shape files, etc. The module offers features for users to create linear combinations of site characteristics into aggregate representations of the overall site suitability to absorb development.
- The allocation mechanism offers users a variety of allocation methods. Land absorption for some sectors, such as employment, is clearly closely related to site characteristics and local markets. Other sectors of interest to MAG, for example prison populations, do not really observe site location characteristics and are better prorated to certain user-defined locations. Yet other sectors, particularly "work-at-home" employment, are best allocated to lands that have already been allocated—in this instance, residential development.
- The trip generation mechanism is programmable. SAM-IM will generate a trip generation data set for any TAZ system of interest. SAM-IM will create trip generation data sets according to any file format defined by the user. SAM-IM will create estimates for "derived" variables according to any set of user-defined equations that are interpreted geographically. (For example, users can automatically generate estimates of population by combining a residential housing shape file with another shape file representing household occupancy rates.)

The Allocation Mechanism in SAM-IM

SAM-IM simulates the growth of regions by looking for land that is best suited for absorbing development.

1. Forecast growth: A set of themes defines the forecasts for each of the allocation sectors for each forecast year. MAG, for example, has created such themes for five-year intervals from the year 2000 to 2040. If the forecasts are from DRAM/EMPAL, SAM-IM disaggregates these from RAZs to TAZs using a RAZ shape file. If the forecasts are known only county wide, as in the case for allocating growth in seasonal and transient populations, then the forecasts are associated with a shape file with one single polygon—the county.
2. Vacant "developable" lands: In order to be considered a candidate to absorb growth, land must be currently undeveloped or planned for redevelopment. An existing land-use shape file, with codes designating "urban vacant" or "agriculture," describe these conditions.
3. "Conforming use": In order to be considered a candidate to absorb growth, land must be declared an appropriate use in the general plan. The shape file describing the General Plan for Maricopa County is used for this. For example, MAG's coding scheme defines five categories of residential use (ranging in density from "rural" to "high density"). All of these uses are considered eligible to absorb growth. SAM-IM will not allocate residential growth to lands dedicated to commercial uses in the general plan, for example.

4. Site suitability scores: Candidate sites then are evaluated with respect to their suitability to absorb growth. This is accomplished through the assignment of a "site suitability score." Site suitability scores are generated countywide, so every piece of land in the county has associated with it a score that reflects its competitive suitability to attract development of certain type.
5. Growth allocation: No matter how derived, site suitability scores drive the allocation. Lands considered to be eligible for development are "tagged for development" in rank-score order, until the growth forecast is fully allocated. Lands that are tagged for development are extracted and moved to a "growth" grid, which accumulates growth associated with each allocation sector, whether it be residential dwelling units, employment, hotel rooms, etc.

At the end of the allocation process, run once for each allocation sector, the growth grid is combined with the then-current existing land use to represent a projection of "future" land use (for the target year in question).

All of these computational operations are not actually carried out on feature themes; instead, land covers and site suitability evaluations are represented as grids. SAM-IM converts them automatically. Grid representation is used for a number of reasons.

- There is no computational geometry necessary. There are no polygon "intersections" or "unions" that have to be constructed.
- The method is entirely "zone-less." Growth is represented by a grid and therefore results can be aggregated upward to any geographic level. Results can be summarized for any TAZ system or for any polygon theme, whether it represent TAZs, RAZs, census tracts, zipcodes, municipal boundaries, water districts, or school districts.
- The method is unbiased with respect to the way that input data sets are coded. The basic allocation "land unit" is a small uniform grid cell.

The planner can set the grid resolution desired for SAM-IM. Currently, because of ArcView addressing limitations, 220-foot cells (1.11 acres) is the smallest resolution that SAM-IM can handle for the 9,000 square mile Maricopa County. Smaller modeling areas, of course, could be modeled at much higher resolution.

The actual implementation of the allocation mechanism in SAM-IM offers a number of other add-ons and features not mentioned above. These include:

- Subarea control totals: A common practice in land-use forecasting is to translate regional forecasts of population and employment into subareas first, before allocating it to smaller geographic units (such as TAZs). Regional Analysis Zones (RAZs), the geographi unit by which DRAM/EMPAL generates forecasts, is a type of subarea. SAM-IM will observe any kind of subarea control totals that the planner wishes to express. The user could even express subareas in terms of "sub-markets," for example distinguishing between "downtown," "airport," and "suburban" office space demand.

Figures 3. SAM-IM predicts future growth and urbanization by simulating development. The allocation mechanism evaluates the suitability of land to absorb development and chooses the most appropriate sites in conformance with general plans.

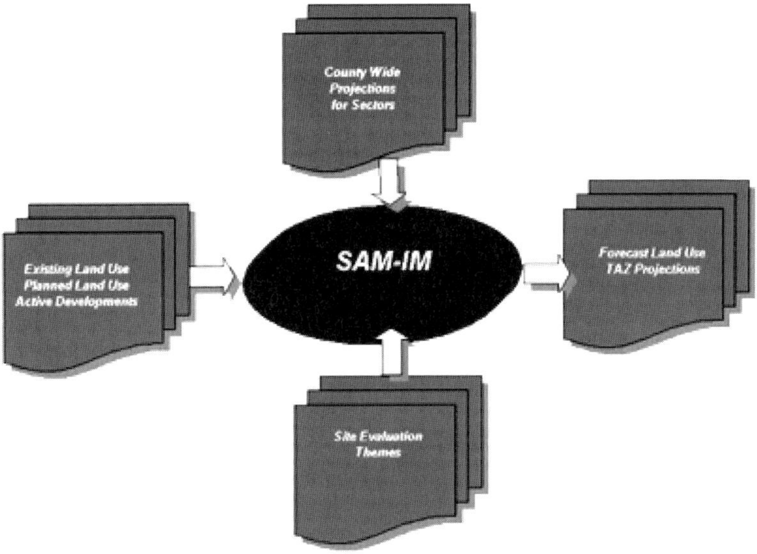

- Development size: The allocation mechanism in SAM-IM involves the "tagging" of grid cells considered to be the most suitable to absorb development. However residential and commercial developments do not usually occur one cell at a time. SAM-IM will "bundle" adjacent cells together during the allocation process so that growth reflects the development of projects consistent with average development sizes recorded for the region.
- Known projects: As a practical matter, planners know where much of the growth forecast for a region will go, especially for the near-term years of a projection series, because these projects already have to be positioned somewhere in the development approval process for them to actually arise. MAG maintains databases tracking development projects throughout the region: their attributes and densities as well as their likely timing. One of the expectations of a forecast is that it should reflect the appearance of projects that are underway, and SAM-IM does this.
- Prorating growth: Site location mechanisms based on the suitability of land to absorb growth are not necessarily the best way to allocate growth for some sectors. Growth in prison population, for example, is such a sector. For some sectors, "prorating" growth among a discrete set of known existing sites is the best approach. SAM-IM provides mechanisms to allocate according to this method.
- Redevelopment overrides: SAM-IM requires land to be "developable" before it considers it eligible to absorb growth. Normally, this land is currently vacant. SAM-IM will, however, allocate to redevelopment areas. SAM-IM provides a mechanism by which formally adopted redevelopment districts are considered eligible for growth and development regardless of the underlying existing uses.

- Prohibition layers: Rivers, flood control channels, aqueducts, and irrigation canals are all hydrological features that are normally reflected in land covers assembled in Maricopa County and therefore SAM-IM can rightfully consider them ineligible for development. Flood plains, however, are an example of land features that are not normally reflected in either existing land-use or planned land-use coverages. Growth prohibitions imposed by municipalities through policy, although not one that currently appears in Maricopa County, is another example. SAM-IM provides two different mechanisms for addressing the suitability of land to absorb development for factors other than strict land use. One is to reflect these characteristics in the site suitability evaluation grids; the other is to declare these areas to be prohibited. SAM-IM lets users create "undevelopable" layers to restrict development for reasons other than land use.
- Reallocation mechanism: A practical reality in modeling land use is that source data acquired from different sources or generated by different models can be inconsistent. That is, SAM-IM can be asked to allocate growth where there is no eligible land, or insufficient land, to absorb it. Regardless, SAM-IM reports these events, collects "residual" growth that could not be allocated, and provides a mechanism for allocating it on some other basis, such as by increasing general plan densities or by relaxing preset control totals.

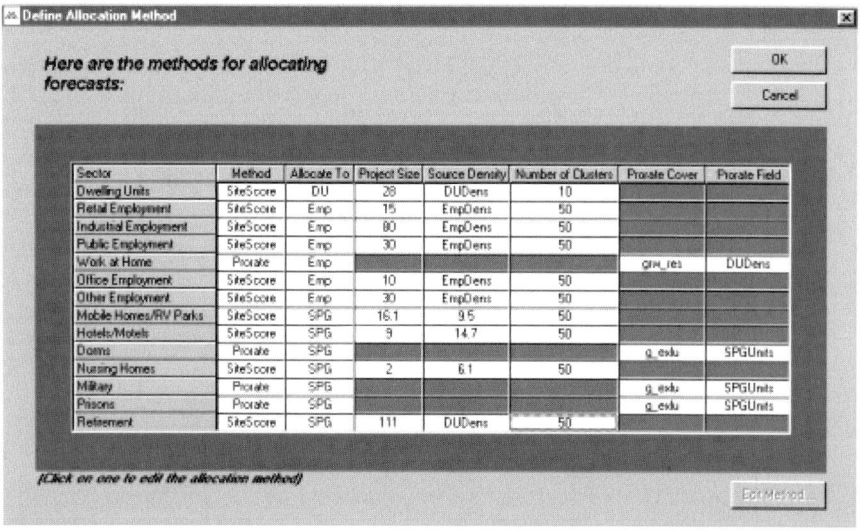

Figure 4. Users can define how allocations should be performed for each of the 14 different forecasting sectors. SAM-IM provides a variety of methods for allocating growth, the most sophisticated of which involves an appraisal of the suitability of land to absorb different kinds of development.

Site Suitability Evaluation

The growth allocation mechanism in SAM-IM is driven by the evaluation of site suitability. Growth is allocated to sites (grid cells) that are considered to be more suitable than their competing neighbors. This is accomplished by representing the evaluation of land to absorb development by "site suitability scores." Growth is allocated to land on a rank-order basis.

The model itself provides this mechanism. It will allocate according to any "site suitability evaluation" system defined for it, linear or nonlinear, normalized or nonnormalized—the only requirement is that it must be expressed as a grid. The modeler, however, can create any kind of site suitability scoring system, reflecting any set of site characteristics or conditions, for which there is information.

The original site evaluation methodology incorporated these measures:

- Highway proximity: Present day and planned transportation networks for the future are used to evaluate land access. They also serve a surrogate for proximity to infrastructure.
- Proximity to urban development: Distance to other "built" uses is another measure that is used as an indicator of the availability of local infrastructure and utilities to support development.
- Neighboring "built" uses: The amount of "built" use in the immediate neighborhood is a measure reflecting infill potential. Infill projects are favored over those in more remote locations.
- Development probability: Planners who were overseeing the progress of projects that have already initiated the development review process assessed the potential and likelihood associated with these developments.

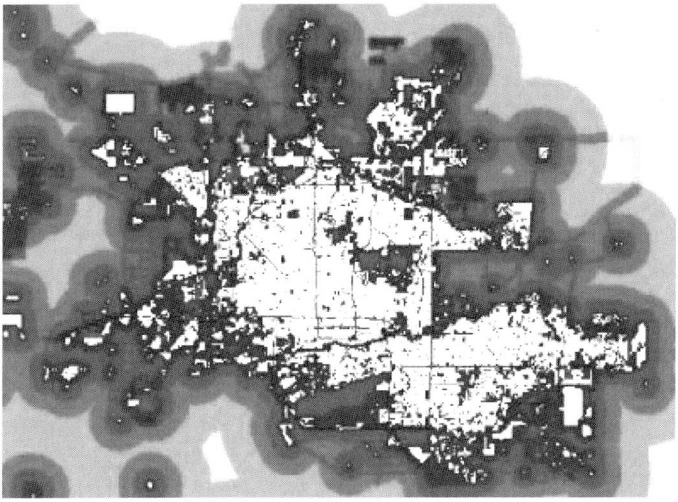

Figure 5. This is a composite image of the overall evaluation of sites for residential land, with darker shades reflecting better locations (in areas not already built). SAM-IM uses this land evaluation theme to target growth.

Calibrating Site Evaluation "Scoring" Equations

The original set of site evaluation measures hypothesized for Maricopa County has not been formally calibrated. However, this calibration is scheduled to occur during the summer of 2000. The calibration procedure will be based on an analysis of land-use change between two different points in time. Development and growth occurring over this time frame can be identified on a cell-by-cell basis, along with various measures of the features and characteristics associated with growth cells.

SAM-IM Trip Generation

One of the principal goals of SAM-IM is to provide forecasts for the 1,549 TAZs required by the transportation model. In addition, though, MAG's transportation model requires TAZ-level forecasts for variables other than those that come from the DRAM/EMPAL and SAM-IM land-use modeling environments. Examples are projections of enrollments at post-high school campuses and total air traffic embarkations at Sky Harbor Airport. Projections of both of these variables are geographic-specific but are prepared completely outside of the land-use forecasting environment.

SAM-IM contains a module that provides features for generating trip generation data sets from SAM-IM forecast land-use covers. These features provide for the estimation of parameters "derived" from the land-use forecasts themselves, as well as for the insertion of forecasts of other independent variables acquired from other independent sources. The module will assemble these various sources, will do the necessary computations, and will generate the results according to any user-specified TAZ geography. Furthermore, SAM-IM will let the user specify the output format of the data.

Historically, TAZ geographies and data requirements change with the evolution of the transportation model. Consequently, it is important for SAM-IM trip generation functions to be programmable.

Similar in concept to ESRI's new "Model Builder" extension (but predating it), SAM-IM allows users to write "programs" that express how the trip generation database is to be created. The "program" has these features:

- Users can express what transportation model variables should be drawn directly from the SAM-IM land-use allocations.
- Users can express equations by which "derived" variables required by the transportation model should be computed.
- Users can identify different sources of computation parameters—for example a shape file that describes "persons-per-household" assumptions for zones.
- Users can identify different sources of forecasts for other independent variables—for example a shape file that records independent forecasts of passenger activity at Sky Harbor Airport.
- Users can define what the trip generation database should look like—its field structure (names and data types) as well as its output format.

Figure 6. The Trip Generation feature of SAM-IM combines variables drawn from the land-use allocation and summarizes them for any TAZ system required by the transportation model. In addition, users can write expressions to generate estimates of any derived variable drawn from factors represented by other GIS covers.

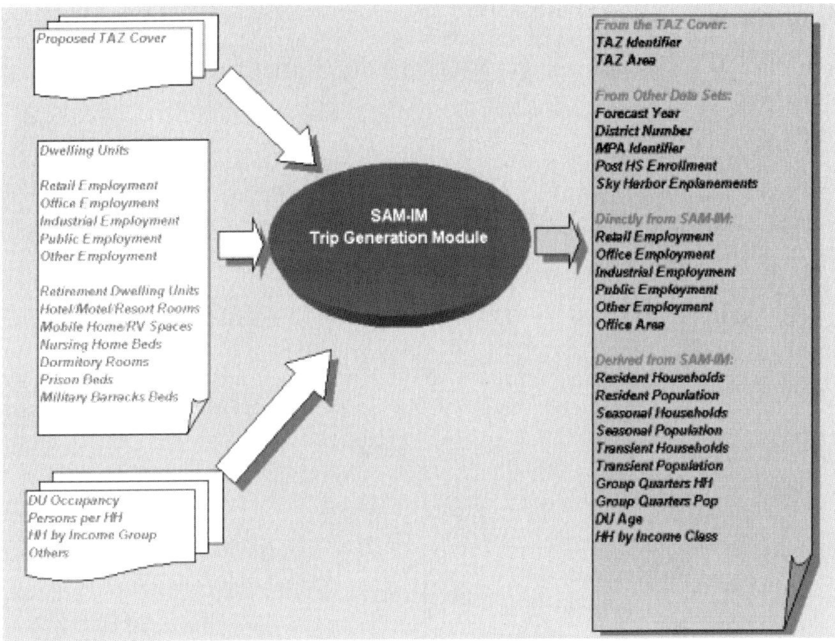

Implementation Issues

Historically, much of the tedious, labor-intensive work with land-use and transportation databases is driven by the need to operate models with inherently different geographical zone structures. Furthermore, much of the source data that is used to generate estimates of many variables comes from yet other geographies.

Zone system database maintenance is one of the major impediments to migrating data sets created from one modeling environment to another, and therefore is a major concern. Historically, much attention was focused on maintaining a variety of geographic zone structures, making sure that zone structures are "upwardly" compatible. Changes made in one zone structure impact all of the others for which compatibility is being maintained. Every change to zone geography carries with it attendant database issues—households, employment, resident population, visitor population, and other attributes all have to be reestimated for the new zone system, often requiring a retreat back to the original source data.

Consequently, one of the principal objectives of SAM-IM was to build a model and a database structure that was fundamentally "zone-less" in nature—where land and development, both now and in the future, was not dependent on any geographic system of zones. Consequently, aggregating land-use information to any zone system that might be needed to drive a land-use or transportation model could be done fully automatically and routinely.

The best way to represent land use at a micro-level and on a "zone-less" system is with grid cells ("raster geography"), such as those inherent to ARC/INFO GRID and the ArcView Spatial Analyst. The finely-grained, most detailed, level of geography typically possible with conventional vector (polygon) representation presumably would be the "parcel" level—but this is not good enough. First, "parcel" coverages are simply too big—processing a parcel database for Maricopa County of over one million records is too much for ArcView to perform effectively. Also, parcels themselves can easily cross zone boundary lines thereby introducing the very same zone-boundary problem that the model is designed to resolve. And finally, land-use development that is being simulated primarily occurs in undeveloped and unsubdivided regions—the exact places that the parcel coverage does not support effectively.

Grid representation also gives the model access to a number of useful spatial analysis functions that do not exist for conventional vector geographies. A proximity calculation, for example, cannot be effectively measured for polygons in an unbiased way because of the variation in size in polygons. Grid representations of geography, on the other hand, can treat land based on its specific cell location, regardless of what polygon it is a part of. Grid representations also offer an entire family of raster functions and operations, such as "proximity," "neighborhood," "partial selection," "compute," and "combine" that simply do not exist for vector geographies.

The SAM-IM allocation mechanism depends heavily on a grid representation of land-use geography. The basic premise of the model would not even be possible without the capabilities of ArcView Spatial Analyst or something comparable. This being said, the implementation of the model using ArcView grid did present us with several technical issues. These are covered next.

Single attribute grid representation

Much of the grid representation of geography offered by the ArcView Spatial Analyst seems to have been driven by natural resource applications, where each grid represents a single specific attribute of land: for example, soil type, permeability, vegetation classification, etc. The data structures needed for urban land-use modeling are much more complex—they need to carry a number of different attributes which a single grid cell can exhibit, including land-use code, dwelling unit density, employment density, and densities associated with any one of a number of different use characteristics (nursing homes, hotel and motel rooms, housing classes, etc.). A "single attribute" representation leads to a proliferation of grid databases and high execution time needed to build and process the databases. Fortunately, ArcView does provide mechanisms to associate multiple attributes with a single cell through the "VAT," which is virtually the same as a standard dBase or INFO table. If land-use grids are built based on a "polygon identification number," then all of the attributes associated with the original polygon can be migrated to the pertinent grid cell and the "VAT" of attributes can be manipulated with conventional database management functions (i.e., fields can be added, deleted, etc.).

Issues with spatial analyst

The ArcView Spatial Analyst proved to be sufficiently stable to support the SAM-IM model, a model that has grown significantly in computational complexity and sophistication. Still, there are several aspects associated with the Spatial Analyst that posed serious problems and limitations. These are:

- The ArcView Spatial Analyst running in Windows environments will not reliably delete grid data sets. This is extremely unfortunate for SAM-IM, because the model generates a large number of "temporary" datasets that must be deleted throughout the process.
- Even though the ArcView Spatial Analyst is a "Windows" program, one of the undocumented aspects of it is that grid names must conform to the old DOS 8.3 file naming convention. The Spatial Analyst offers no warning about this—failure to observe the DOS 8.3 naming convention causes ArcView to behave erratically.
- The ArcView Spatial Analyst imposes serious limitations on the sizes of data sets that can be built. Reportedly, the size limitation is the 32-bit addressing barrier (approximately 2.5 gb). It appears that some Spatial Analyst functions, particularly "grid combine," generate temporary grid data sets much larger than are ultimately necessary.

SAM-IM contains large bodies of code to avoid some of these limitations. Some, however, can not be addressed in application code and therefore have become limitations and constraints on the model itself.

For example, SAM-IM users are limited when designing site selection scoring equations or reflecting the results of a statistical calibration. Our practical experiences at MAG are that site scores have to be "normalized" within certain restricted numeric ranges, such as integers between 0 and 100. This is not an absolute rule, but one that applies specifically to Maricopa County—the size of the grid cell used there, along with the number of land-use and general plan polygons, all conspire to fail due to size restrictions if the scoring theme has more than 100 individual scores. The limitation can not be determined in advance—it can only be determined by trial.

Variable grid cell sizes

MAG's legal jurisdiction for regional land-use, transportation, and air quality modeling covers all of Maricopa County. Maricopa County, like other planning areas in the Southwestern United States, is large (9,000+ square miles). And therefore, it is appropriate that the scope of the modeling area should match it even though a large part of land in Maricopa County is undeveloped desert.

The ArcView Spatial Analyst requires grids to be uniform with respect to grid cell size. This seems entirely natural (compute screens don't have pixels of variable dimensions), but for land-use modeling it would be helpful if grid cells could be variable in size. The size of grid cell sizes for representing land use such as SAM-IM is driven by the need for resolution.

SAM-IM currently runs at grid cell sizes of 220 feet on a side, or approximately 1.11 acres. This appears to push ArcView to the very limit of its constraints on grid sizes and yet is insufficient for densely developed areas, such as downtowns. There are highly developed uses in downtowns, such as a 20-story building on a half-acre parcel, that literally "disappear" during the conversion from vector representation (the land-use polygon cover) to grid. A mechanism to exchange higher resolution in densely developed areas, such as downtowns, for lower resolution in undeveloped deserts and forests, would be helpful.

Loss of information during database conversions

The normal representation of urban land use is the conventional vector feature database of polygons—each representing "generalized" land use (that is, a "shape file"). Land-use data is typically digitized this way and is maintained this way. Therefore, SAM-IM automatically converts land-use data from its normal vector form to grids in order to perform the growth simulation. Once accomplished, the resulting "growth" grid is converted back to its vector form.

Land-use modelers should expect, though, that there is a natural loss in information associated with land use through the conversion process. Anyone who has spent time looking at aerial photographs knows that you can see things in aerial photomosaics shot at submeter resolution (e.g., buildings) that you can not see in imagery shot at higher resolutions (e.g., 30-meter satellite images). The same holds true for grid representations—the simple conversion of a land use "shape file" to grid and back again will not give you the same polygon representation that you started with, except at very high resolutions.

The problems are:

- First, it is completely possible that entire land-use polygons will "drop out" of the grid representation during the conversion process. It is easy to see that quarter-acre parcels, for example, can not be completely captured by four-acre grid cells. As we mentioned earlier, this is an issue that arises in downtowns, in particular. Unfortunately, in urban land-use modeling it is completely unacceptable to "lose" 15,000 employees that work in a single high-rise office building.
- Also, "area" is not preserved. For example, it is entirely likely that a 40-acre rectangular parcel to be represented by 44 (or some other number) one-acre grid cells. So the "area" of the gridded polygon is 44 acres, not the 40 acres associated with the original polygon from which they came. The error associated with grid areas is not consistent—it all depends on the resolution of the grid cell and the location of the polygon within the grid.

Future Developments

During the next phase of work, a number of enhancements are planned. These are:

- Data enhancement project: SAM-IM makes use of land coverages that describe existing land use, planned land use, planned development project, active development projects, as well as transportation networks from the forecasting model and other coverages describing retirement communities and redevelopment communities. These databases all have been assembled from a variety of disparate sources. One of the issues influencing the performance of the forecasting series (DRAM/EMPAL and EMME/2, as well as SAM-IM) has been the consistency of information portrayed among these datasets. MAG's Data Enhancement Project, begun in early 2000, is a comprehensive database enhancement project aimed at updating all of these databases and bringing them into concurrence.
- Mixed-use developments: Historically, the land-use data models embedded in SAM-IM have presumed that land is always "dedicated" to a single use, whether it be residential, commercial, industrial, or some type of unbuilt use. Consequently, SAM-IM does not have an effective way to address mixed use developments, without converting these to single dedicated uses. A future enhancement target is to fully support mixed use developments, with variable percentages of other land uses.
- Densification mechanism: The allocation mechanism in the SAM family of models maintained target development densities defined in the general plan. In fact, general plans are often ambiguous about precise development densities, which can be higher than the target densities in the general plan while still being consistent with the plan. Additionally, densification does tend to increase as the supply of developable land decreases. One of the enhancement areas targeted for SAM-IM will provide the ability to increase development densities associated with planned land uses.
- Development phasing: MAG tracks development activity carefully and has assembled data bases that describe these projects as they unfold, including information about phased approvals of developments. MAG has developed a "development velocity" concept by which the phased development activity can be projected in future years. The "development velocity" concept will be automated in the next version.
- Redevelopment Activity: As indicated earlier, SAM-IM allocates growth to lands that are defined to be eligible to absorb development. Ordinarily, these lands are vacant or are "unbuilt" in some sense, except in "redevelopment" areas. In formally adopted redevelopment districts, SAM-IM considers all land, built or otherwise, to be eligible for "next-generation development." The next version will have a mechanism to redistribute uses that are replaced.

Current and Future Applications

SAM-IM is becoming an integral part of the forecasting process at MAG. The predecessor model SAM played a principal role in the development of the official population and employment projections by TAZ which were adopted by the MAG Regional Council in June 1997. Data being collected by the 2000 Census and the ongoing MAG Database Enhancement Project will be input to SAM-IM to prepare the next set of TAZ projections for Maricopa County during 2002.

In the meantime MAG is testing a number of land-use development scenarios. For example, during the early part of 2000 MAG applied SAM-IM to produce alternative population and employment allocations for:

- a light rail corridor study, based on updated general plans and known developments
- no-build transportation scenarios required in air quality conformity analyses
- out-year (2040) trip generation data consistent with the adopted TAZ projections to 2020
- alternative highway alignments in a subregion which is growing rapidly

In these tests, SAM-IM demonstrated the capability to provide reasonable TAZ-level estimates of population and employment, given updated GIS coverages of existing land use, general plan land use, and/or known development projects. It is anticipated that the model will be used to generate additional land-use alternatives during the next year and that these "tests" will disclose areas of needed improvement. SAM-IM will continue to be calibrated and refined in preparation for the next official MAG forecasting cycle in 2002.

Because the model has proven to be both useful and configurable, next year MAG intends to provide SAM-IM, with appropriate manuals and training, to its member agencies, i.e., 24 cities and towns and the county, many of which already use ArcView. It is anticipated that SAM-IM will be used by these agencies to evaluate the impact of local development policies and plans. Some may choose to focus the model within their geographical boundaries and/or use smaller spatial units than TAZs. SAM-IM's structure will allow users to tailor the modeling domain, grid-size, and aggregation of model output to local land-use planning needs.

MAG's experience with SAM over the last five years indicates that the model's greatest asset may be its flexibility. It can be used to prepare socioeconomic forecasts, as well as to analyze alternative development patterns, perform what-if analyses, at varying spatial scales.

Due to concerns about the impacts of growth on infrastructure such as roads, utilities, and schools, and the overall quality of life, the Arizona legislature has recently passed "growing smarter" laws, requiring jurisdictions to develop new land-use plans. This increased emphasis on accurate and potentially-binding plans, along with the looming possibility of growth boundaries, is sparking an interest in land-use models. Fortunately, MAG and its member agencies are well-positioned to address future growth and other regional and local land-use issues, thanks to SAM-IM.

Contributor Biographies

Anubhav Bagley is the Socioeconomic Modeling Program manager at the Maricopa Association of Governments (MAG). He is responsible for the development and use of MAG socioeconomic models. Mr. Bagley has five years of professional experience in geographic information systems (GIS), land-use planning, socioeconomic analysis and modeling, environmental impact analysis, and site assessments and evaluation. Prior to joining MAG, Mr. Bagley worked with the environmental and transportation groups of Science Applications International Corporation. Anubhav has a master of environmental planning from Arizona State University and bachelor in planning from School of Planning and Architecture, New Delhi, India.

Tasila Banda-Sakala is currently working on a doctoral degree at the University of California, Davis. Her research is being conducted in Tanzania.

Michael Batty is professor of spatial analysis and planning, and director of the Centre for Advanced Spatial Analysis (CASA) at University College London (UCL). He holds a joint appointment between the Bartlett School of Architecture and the Department of Geography. He has made many contributions to the development of computer models of cities and regions and recent work is focused on dynamic models of urban development and the visualization of cities using virtual reality methods. His books range from *Urban Modelling* to *Fractal Cities*. He is editor of *Environment and Planning B* and a Fellow of the British Academy. The work of his group can be seen at www.casa.ucl.ac.uk/.

Ward Brady is professor and program leader for the Environmental Resources Program in the Morrison School at Arizona State University East. His research interests include vegetation and landscape ecology with an emphasis on quantitative methods and monitoring.

Betsy Conklin is completing her master of science degree at Arizona State University East. Her research is concerned with the temporal and spatial distribution of TCE in the aquifers underlying the Phoenix metropolitan area.

John David obtained his master's degree from the Department of Plant Biology of Arizona State University under the guidance of Dr. Jianguo Wu in 2001. He has extensive experience in computer programming and restoration ecology. His thesis work focused on the development of the hierarchical patch dynamics modeling platform.

Philip C. Emmi is professor of urban and regional planning at the University of Utah where he serves as director of the Urban Planning Program and of the interdisciplinary graduate certificate program in the adaptive management of environmental systems. He holds degrees from Harvard University and the University of North Carolina at Chapel Hill. He has been a Peace Corps volunteer, a Fulbright Scholar, Lowell Bennion Community Service Professor, and a National Science Foundation grant recipient. His academic interests include planning theory, cities and sustainability, and the use of dynamic simulation modeling for collaborative learning in contexts of competition and conflict.

Craig Forster is research associate professor in the Department of Geology & Geophysics at the University of Utah. Originally a hydrogeologist, he spent more than two decades successfully studying and modeling fluid flow and mass/heat transport through complex geologic systems. He has coauthored more than 30 peer-reviewed papers in this arena. Dr. Forster now maps and models the complexities, linkages, and feedbacks found at the interface between social institutions and the natural environment. Dr. Forster contributes to and leads interdisciplinary teams that build system dynamics models addressing the consequences of population growth and climate variability in urban communities and human-affected ecosystems.

William E. Grant has been teaching a graduate course on the use of systems analysis and simulation in ecology and natural resource management in the Department of Wildlife and Fisheries Sciences at Texas A&M University since 1976. He has authored three textbooks on this topic and has published more than 130 articles in scientific journals. He has served as a member of the board of governors and as president of the International Society of Ecological Modelling (ISEM) and as a member of the Editorial Advisory Board of the international journal *Ecological Modelling*, of which he is currently associate editor.

Subhrajit Guhathakurta is an associate professor in the School of Planning and Landscape Architecture at Arizona State University. He received a Ph.D. in city and regional planning from the University of California, Berkeley, and a master's degree in community and regional planning from Iowa State University. His research and publications have spanned a range of subjects including land-use and environmental modeling, economic development planning, and housing. His paper on urban modeling and planning theory received the Chester Rapkin Award for best article of the year in volume 18 of *Journal of Planning Education and Research*.

Lewis D. Hopkins, FAICP, is professor in urban and regional planning and associate dean of the College of Fine and Applied Arts at the University of Illinois at Urbana-Champaign. He was head of the department for 13 years, editor of the *Journal of Planning Education and Research*, Fulbright Senior Scholar in Nepal, and chair of the Planning Accreditation Board. He earned a bachelor of arts in architecture, master of regional planning, and Ph.D. in city planning from the University of Pennsylvania. His recent book, *Urban Development: The Logic of Making Plans* explains how plans work and when they are likely to be made and worth making. His current work focuses on computer interfaces that access the logic in plans and enable collaboration. He serves on the Urbana Planning Commission.

G. Darrel Jenerette is completing his Ph.D. in the Department of Plant Biology, Arizona State University under the guidance of Dr. Jianguo Wu. He is primarily interested in understanding ecological complexity and self-organization, in particular the factors affecting the generation of order and the functional consequences of ecological organization. This theoretical basis underpins his research in areas of landscape ecology, biogeochemistry, and urban ecology. His dissertation research addresses the multiple-scale spatial patterns of soil carbon and nitrogen in the greater Phoenix, Arizona region. He is also engaged in understanding ecological and sociological interactions via city acquisition of renewable freshwater.

John D. Landis is the chair of the Department of City and Regional Planning at the University of California Berkeley where he teaches graduate courses in planning history and methods, land-use planning, project development, and GIS. His recent research has focused on the causes and consequences of urban sprawl, affordable housing, and statewide land-use planning policy and planning.

Laura Musacchio is assistant professor in the School of Planning and Landscape Architecture and Center for Environmental Studies at Arizona State University. Her research includes the development of knowledge about the human dimensions of landscape ecology and urban ecology. Her current investigations focus on the modeling of the dynamics of planned and designed landscapes as self-organized systems within an ecoregional context. In her landscape models, she focuses on how human decision making such as those made in the design and planning processes, can affect the spatial and functional heterogeneity of urban patterns and how changes in these patterns affect ecosystem health and services such as water quality, wildlife habitat quality, visual quality, and recreational access quality. Her current research projects include the Rio Alamar Urban River Restoration Project in Tijuana, Mexico, and landscape change of suburbanizing floodplains and watersheds in the Phoenix metropolitan region with the Central Arizona-Phoenix Long-Term Ecological Research Project in urban ecology (CAPLTER).

Tarla Rai Peterson is associate professor of communication at the University of Utah. Her research in environmental communication focuses on systems analysis of environmental conflict and sustainable development. She also examines the relationship between dissent and political deliberation in democracy. Most of her research is conducted as part of interdisciplinary teams funded by various sources, including the National Science Foundation, the Texas Agricultural Extension Service, and the Environmental Protection Agency. She has consulted widely both in the United States and Europe. She has authored or coauthored over 50 articles and book chapters, as well as a book titled, *Sharing the Earth: The Rhetoric of Sustainable Development.*

Micahle K. Reilly is a Ph.D. student in the Department of City and Regional Planning at the University of California Berkeley where he teaches graduate courses in spatial analysis and GIS. His current research focuses on the effects of how the urban environment is perceived and experienced on travel behavior.

Qing Shen is an associate professor of urban planning at the University of Maryland, College Park. His fields of research and teaching are metropolitan planning, urban spatial structure, and statistical methods and GIS applications. He is the author of numerous publications in refereed journals. Dr. Shen was born in the People's Republic of China. He was educated in China (Zhejiang University) and Canada (University of British Columbia) before coming to the United States. He holds a Ph.D. in city and regional planning from University of California, Berkeley. He previously was an associate professor of urban studies and planning at MIT.

C. Scott Smith is an Arizona State University School of Planning and Landscape Architecture alumnus and former comanager of ASU's interdisciplinary Spatial Research Support Lab. His research addresses issues concerning social equity, transportation planning, and GIScience. He is currently pursuing a Ph.D. in planning, policy, and design at University of California, Irvine's School of Social Ecology.

Rita Walton is the information services manager at the Maricopa Association of Governments (MAG). She is responsible for socioeconomic analysis and modeling and telecommunications and automation support. Ms. Walton heads up the Maricopa Association of Governments Information Center (MAGIC), established as a subsidiary of MAG in June 1998, to provide member agencies and the community with regional information. Prior to joining MAG, Ms. Walton was vice president of development and client support for an occupational health and safety software company; manager, west region economic consulting practice at Coopers and Lybrand; and vice president of urban and regional analysis at Mountain West, a local real estate consulting company.

Gary Whysong received his bachelor of science and master of science degrees from Montana State University and Ph.D. from the University of Wyoming. After four years of teaching and research in Alberta, Canada, he joined the Environmental Resources program at Arizona State University. Currently, he teaches in the areas of statistics, vegetation and animal measurements, and computer simulation modeling. His research interests are in the application of GIS/remote sensing and computer simulation modeling to natural resource management.

Jianguo Wu is associate professor in Department of Plant Biology, Arizona State University. He obtained his bachelor of science from Inner Mongolia University and master of science and Ph.D. from Miami University, Oxford, Ohio. His research focuses on landscape ecology, urban ecology, and hierarchical patch dynamics. He has published two books and over 90 journal papers and book chapters. He serves on editorial boards of *Landscape Ecology, Geographic Information Sciences, Acta Ecologica Sinica,* and *Acta Phytoecologica Sinica.* He has been program chair of the U.S. Association of Landscape Ecology (US-IALE 2001) and chair of Asian Ecology Section of Ecological Society of America (1999–2000).

Wei-Ning Xiang is a professor of geography and earth sciences at University of North Carolina at Charlotte. A native of Beijing, China, he came to the United States in January 1985. He earned his master's degree in regional planning from University of Massachusetts at Amherst in 1986 and doctoral degree in city and regional planning from University of California at Berkeley in 1989. His academic interests are in the areas of geographic information systems (GIS), environmental analysis and land-use planning, and spatial decision support systems (SDSS). He has published more than 20 refereed journal articles and is on the editorial board of the international journal *Environment and Planning B: Planning and Design.*